# All the Multiverse!
# II
## Between Multiverse Universes

*Quantum Entanglement Explained by the Multiverse*
*Coherent Baryonic Radiation Devices - PHASERs*
*Neutron Star Multiverse Slingshot Dynamics*
*Spiritual and UFO Events, and the Multiverse*
*Microscopic Entry into the Multiverse*

### Volume 2: All the Multiverse!

*When Man has conquered all the depths of space and the mysteries of time,*
*Still he will be at the beginning.*
*Things to Come - H. G. Wells*

# All the Multiverse!
## II
### Between Multiverse Universes

*Quantum Entanglement Explained by the Multiverse*
*Coherent Baryonic Radiation Devices - PHASERs*
*Neutron Star Multiverse Slingshot Dynamics*
*Spiritual and UFO Events, and the Multiverse*
*Microscopic Entry into the Multiverse*

Volume 2: All the Multiverse!

## STEPHEN BLAHA

## BLAHA RESEARCH

**Cover Credits**

A symbolic representation of the Multiverse, with a recent picture of the author portrayed as looking into our universe from the outer multiverse. Dark spots represent universes. Copyright © 2014 by Stephen Blaha. All Rights Reserved.

Rev. 00/00/01        July 9, 2014

*To My Grandchildren:*

*Nicholas, Maxim, Alexandre, and Milan*

## Some Other Books by Stephen Blaha

*All the Multiverse! Starships Exploring the Endless Universes of the Cosmos using the Baryonic Force* (Blaha Research, Auburn, NH, 2014)

*All the Universe! Faster Than Light Tachyon Quark Starships & Particle Accelerators with the LHC as a Prototype Starship Drive Scientific Edition* (Pingree-Hill Publishing, Auburn, NH, 2011).

*From Asynchronous Logic to The Standard Model to Superflight to the Stars* (Blaha Research, Auburn, NH, 2011)

*From Asynchronous Logic to The Standard Model to Superflight to the Stars; Volume 2: Superluminal CP and CPT, U(4) Complex General Relativity and The Standard Model, Complex Vierbein General Relativity, Kinetic Theory, Thermodynamics* (Blaha Research, Auburn, NH, 2012)

*The Algebra of Thought & Reality: The Mathematical Basis for Plato's Theory of Ideas, and Reality Extended to Include A Priori Observers and Space-Time; Second Edition* (Pingree-Hill Publishing, Auburn, NH, 2009)

*Quantum Big Bang Cosmology: Complex Space-time General Relativity, Quantum Coordinates, Dodecahedral Universe, Inflation, and New Spin 0, ½, 1 & 2 Tachyons & Imagyons™* (Pingree-Hill Publishing, Auburn, NH, 2004)

*SuperCivilizations: Civilizations as Superorganisms* (McMann-Fisher Publishing, Auburn, NH, 2010)

*Standard Model Symmetries, And Four and Sixteen Dimension Complex Relativity; The Origin Of Higgs Mass Terms* (Blaha Research, Auburn, NH, 2012)

*The Bridge to Dark Matter; A New Sister Universe; Dark Energy; Inflatons; Quantum Big Bang; Superluminal Physics; An Extended Standard Model Based on Geometry* (Blaha Research, Auburn, NH, 2013)

*Universes and Multiverses: From a New Standard Model to a Physical Multiverse; The Big Bang; Our Sister Universe's Wormhole; Origin of the Cosmological Constant, Spatial Asymmetry of the Universe, and its Web of Galaxies; A Baryonic Field between Universes and Particles; Flatverse Extended Wheeler-DeWitt Equation* (Blaha Research, Auburn, NH, 2014)

Available on bn.com, Amazon.com, Amazon.co.uk and other international web sites as well as at better bookstores (through Ingram Distributors).

# Preface

In volume 1 of *All the Universe!* we described details of the exploration of the multiverse – the flat 16-dimensional space that contains a multitude of universes. We described details of a type of starship called a uniship that could escape from our universe and travel to other universes. The discussion was based on an Extended Standard Model of Elementary Particles and an extended Quantum Gravity – both based on the four-dimensional space-time with complex valued coordinates. We also shown that a *complex* sixteen-dimensional multiverse consisting of universes embedded in a flat space, with each universe satisfying a complex Wheeler-Dewitt equation, leads to an understanding of many features seen in our universe: the origin of the Cosmological Constant, the spatial asymmetry of our universe, and the web of Galaxies recently found in our universe. It also suggested a wormhole connection to a sister universe and the existence of a baryonic field between universes and particles (Blaha (2014a)). *The combination of these successes in explaining both the microscopic and cosmic features of our universe, and resolving quantum issues of the Wheeler-DeWitt equation, strongly suggested our theory was in accord with Nature. It gave us a reason to look into the distant future with confidence.*

In this volume we look particularly at the slingshot mechanism using neutron stars to propel starships into the multiverse. The existence of the multiverse implies every neighborhood of every point in our universe contains points of the multiverse. In particular, these multiverse points are points in the Flatverse – the flat 16-dimensional space between universes. Thus we will consider infinitesimal entry into the Flatverse from a point in the universe as well as large scale entry into the Flatverse by uniships using the slingshot mechanism.

Both volumes assume the existence of a baryonic force (between baryons such as nucleons) that leads to baryon number conservation. We perform a preliminary calculation of this force based on deviations in measurements of the gravitational constant G, and suggest a precise experiment to determine the baryonic force coupling constant.

We examine the possibility of making a compact device that would create a strong coherent baryonic gauge field, a PHASER, for microscopic entry into the Flatverse, the possibility of "jumps" in space, the strong possibility of a relation to quantum mechanical entanglement due to the embedding of our universe in the Flatverse, and a seemingly improbable connection to spirits and the "Spirit World" due to the embedding of our universe in the Flatverse.

We also consider the possible types of slingshot mechanisms – especially an umbrella uniship capable of travel in any of the fifteen spatial Flatverse dimensions. We investigate the space-time within a uniship after it emerges from our universe.

This book addresses topics that will be of significance after perhaps tens of thousands of years of research and development – perhaps 50,000 years (four times the approximately 12,500 year period from human hunter-gatherer clans to the present.)

The uniship features that we will describe have many radically different mechanisms. It does not resemble science fiction depictions of starships that are seen in popular movies and television shows. Volume 1 and this volume portray as much as we much as we can forecast about starships and uniships. The descriptions are based on reasonable projections of Physics and certain required assumptions that are reasonable based on our work on the Standard Model, General Relativity, and complex-valued spatial coordinates both in our universe and in the multiverse in general.

The presentation in this book is largely self-contained. The reader may wish to also read some recent books by the author of special relevance for the present work:

*All the Multiverse! Starships Exploring the Endless Universes of the Cosmos using the Baryonic Force* (Volume 1).
*Universes and Multiverses: From a New Standard Model to a Physical Multiverse; The Big Bang; Our Sister Universe's Wormhole; Origin of the Cosmological Constant, Spatial Asymmetry of the Universe, and its Web of Galaxies; A Baryonic Field between Universes and Particles; Flatverse Extended Wheeler-DeWitt Equation.*
*The Bridge to Dark Matter; A New Sister Universe; Dark Energy; Inflatons; Quantum Big Bang; Superluminal Physics; An Extended Standard Model Based on Geometry.*
*From Asynchronous Logic to The Standard Model to Superflight to the Stars; Volume 2: Superluminal CP and CPT, U(4) Complex General Relativity and The Standard Model, Complex Vierbein General Relativity, Kinetic Theory, Thermodynamics.*
*From Asynchronous Logic to The Standard Model to Superflight to the Stars (Volume 1).*
*Standard Model Symmetries, and Four and Sixteen Dimension Complex Relativity; the Origin of Higgs Mass Terms.*

.

# CONTENTS

# FIGURES and TABLES

# 1

# The Multiverse and the Fifth Force: Baryonic Interactions

This chapter summarizes some important points presented in volume I of *All the Multiverse!* and in Blaha (2014a).

## 1.1 Theoretical Evidence for the Multiverse and for Baryonic Interactions

Theoretical evidence for a multiverse of universes is based on the Copenhagen interpretation of quantum theory applied to quantum gravity:[1]

1. Quantum theory requires an "outside" observer to make a measurement of a quantum system. Since we believe our universe is a quantum entity then an observer outside our universe must exist in the Copenhagen interpretation. Thus at least one "outside" universe must exist to play the role of observer. If one additional universe exists it is reasonable to assume that many outside universes exist — the multiverse. These universes must have spatial separation. Since there is no gravitational attraction between universes (Universes have zero total energy.) the space between universes must be flat.

2. There must be a "clock" that sets the time in our universe. This clock must be independent of the contents of the universe. Again we are led to see a need for at least one external universe that acts as a clock. Thus the multiverse.

---

[1] DeWitt, B. S., Phys. Rev. **160**, 1113 (1987); Unruh, W. G., Phys. Rev. D **40**, 1053 (1989) and references therein.

Another justification for a multiverse is that it resolves the Mach paradox. Many attempts have been made to justify the central role of inertial reference frames in General Relativity and the origin of inertia. The existence of flat space outside the universe (the Flatverse) in the multiverse provides a direct justification for inertial reference frames and inertia. Physical Flatverse reference frames are all inertial reference frames.

There are two significant justifications for an interaction between baryons (the fifth force):

1. Baryon number[2] is a conserved quantum number as far as we know. No deviations from baryon conservation have been found. Baryon conservation (like electric charge conservation) can be based on a gauge field similar to the electromagnetic field. However the baryonic field is a 16-dimensional field in the 16-dimensional multiverse in our multiverse theory. (See Blaha (2014a) for details.) Baryonic fields exist between baryons within a universe, and between baryons residing in different universes. It also is a force between the sum total of baryons in differing universes due to its additivity.

2. Like the electromagnetic field it can be used to make quantum measurements on universes. It thus is an integral part of quantum gravity if one requires the Copenhagen interpretation as described above. A quantum gravity observer uses the baryonic field to observe. Without a field between universes, there can be no quantum observations (measurements).

We therefore have a consistent quantum gravity in multiverse universes and a baryonic interaction for quantum measurements between universes.

## 1.2 Experimental Evidence for Baryonic Interactions

The possibility of a fifth force between baryons has been a subject of discussion for over sixty years. The detection of this force experimentally has been complicated by its extreme weakness and its overlap with the force of gravity. The overlap is evident from the total potential[3] between two

---

[2] Baryons are quarks and particles constructed from quarks. The baryon number of a sample is the number of baryons minus the number of antibaryons in the sample.

[3] Blaha (2014b) p. 44.

electromagnetically neutral masses of mass $M_1$ and $M_2$, with baryon numbers $N_1$ and $N_2$

$$V_{tot} = -GM_1M_2/r + (\beta^2/4\pi) N_1N_2/r \qquad (1.1)$$

where G is the gravitational constant, and $\beta$ is analogous to the electric charge e in the electromagnetic Coulomb potential. If both masses are composed of the same substance and have the same mass, then we can set $M_1 = M_2 = M = Nm$ where m is the average mass of the baryons in the masses.[4] In addition we can set $N_1 = N_2 = N$. Then in this case eq. 1.1 becomes

$$V = [-Gm^2 + (\beta^2/4\pi)]N^2/r \qquad (1.2)$$

Note that the gravitational potential term is attractive, and the baryonic potential term is repulsive between baryons.

In considering eqns. 1.1 and 1.2 we have approximated the baryonic potential with only our universe's spatial coordinates. In reality we should be using the spatial separation in Flatverse coordinates. However since our universe is close to flat, the distance between two objects that are not too far apart is approximately the same in both coordinate systems. The baryonic potential in Flatverse coordinates is actually

$$\phi(y_1, y_2, \ldots, y_{15}) = (\beta^2/4\pi)N_1N_2/(y_1^2 + y_2^2 + \ldots + y_{15}^2)^{1/2} \qquad (1.3)$$

Measurements of the gravitational constant G in experiments over the years are significantly different[5,6] – much beyond the estimated measurement errors. The reason(s) for these discrepancies is not known.

We suggest that the difference between these measurements is due to the baryonic force. Thus both the 2010 and 2013 measurements are experimentally correct but disagree because the baryonic force term in eqn. 1.2 creates a difference in effective G values since the experiments used different masses and thus different baryon numbers. Quinn et al found a value for the gravitational constant of $G_1 = 6.67545 \times 10^{-11}$ $m^3kg^{-1}s^{-2}$. The combined 2010 CODATA value for the gravitational constant was $G_2 = 6.67384 \times 10^{-11}$ $m^3kg^{-1}s^{-2}$.

---

[4] We neglect lepton masses since they are negligible relative to the baryon masses.
[5] T. Quinn et al, Phys. Rev. Lett. **111**, 101102 (2013).
[6] P. J. Mohr, B.N. Taylor, and D. B. Newell, Rev. Mod. Phys. 84, 1527 (2012).

These values disagree beyond error bounds. We believe this is due to the fifth force. Thus there is circumstantial evidence for the existence of a fifth force – the baryonic force.

## 1.3 Strength of the Fifth Force - Baryonic Force Coupling Constant

We can make an order of magnitude estimate of the baryonic fine structure constant $\beta^2/4\pi$ using the measured values of the previous section.

Suppose these values are correct and due to a difference in the chemical composition (metals) of the test masses used in the experiment. Quinn et al use 1.2 kg test masses composed of Cu-0.7% Te free machining alloy. The CODATA value being a composite of many experiments does not have an effective equivalent test mass value or composition specified.[7] Suppose the test mass value is $N_1^2 m_1^2 + N_{1e}^2 m_e^2$ for one G measurement giving the result $G_1$

$$-(N_1^2 m_1^2 + N_{1e}^2 m_e^2)G_1 = [-G(m_1^2 N_1^2 + N_{1e}^2 m_e^2) + (\beta^2/4\pi)N_1^2] \quad (1.4)$$

where G is the real value of the gravitational constant. The total test mass is $(m_1^2 N_1^2 + N_{1e}^2 m_e^2)$ with $N_1$ baryons of average mass m in each test mass and $N_{1e}$ leptons of average mass $m_e$.

Suppose further the test mass value is $N_2^2 m_2^2 + N_{2e}^2 m_e^2$ for the other G measurement giving the value $G_2$

$$-(N_2^2 m_2^2 + N_{2e}^2 m_e^2)G_2 = [-G(m_2^2 N_2^2 + N_{2e}^2 m_e^2) + (\beta^2/4\pi)N_2^2] \quad (1.5)$$

where G is the real value of the gravitational constant. The total test mass is $(m_2^2 N_2^2 + N_{2e}^2 m_e^2)$ with $N_2$ baryons of average mass $m_2$ in each test mass and $N_{2e}$ leptons of average mass $m_e$. Since the test masses are electrically neutral and there are approximately equal numbers of protons and neutrons in a test mass it follows approximately that

$$N_{1e} = \tfrac{1}{2}N_1 \quad \text{and} \quad N_{2e} = \tfrac{1}{2}N_2 \quad (1.6)$$

---

[7] The Eötvös' experiment used a 0.1 gm test mass of $RaBr_2$. R. v. Eötvös, D. Pekár, E. Fekete, Annalen der Physik (Leipzig) 68, 11, 1922.

Subtracting eqn. 1.4 from eqn. 1.5 after some algebra[8] we find

$$\Delta G = -G_2 + G_1 = (\beta^2/4\pi)/(m_2^2 + m_e^2/2) - (\beta^2/4\pi)/(m_1^2 + m_e^2/2)$$
$$\simeq (\beta^2/4\pi)(1/m_2^2 - 1/m_1^2) \tag{1.7}$$

The masses $m_1$ and $m_2$ can differ. For example, if $m_H$ is mass of the hydrogen atom, then $m^{-1} = 1m_H^{-1}$ for hydrogen, for carbon $m^{-1} = 1.00782m_H^{-1}$, for copper $m^{-1} = 1.00895m_H^{-1}$, and for lead $m^{-1} = 1.00794m_H^{-1}$.[9] Thus using the Quinn et al and CODATA results and assuming copper and lead test masses we find the order of magnitude *estimate*:

$$\alpha_B = \beta_B^2/4\pi \simeq \Delta G/[(1.00895^2 - 1.00794^2)\, m_H^2]$$
$$\simeq (\Delta G/G)\, G\, m_H^2/.002037$$
$$\simeq (0.000241/0.002037)Gm_H^2$$
$$\simeq .118\, Gm_H^2 \tag{1.8}$$
$$\simeq 5.9 \times 10^{-39} \tag{1.9}$$

giving a very weak baryonic force consistent with our general view of the multiverse. The baryon fine structure constant is minute in comparison to the electromagnetic fine structure constant $\alpha \simeq 1/137$.

## 1.4 The Needed Laboratory Experiment for the Determination of the Baryonic Coupling Constant

The assumed values of $G_1$ and $G_2$ make the calculated value of $\alpha_B$ an order of magnitude estimate since the composition and masses of the test masses in the experiments are not comparable. We suggest that an experimental group measure G with specific differing test masses in the same apparatus to obtain a better value for $\alpha_B$.

---

[8] The reduction of the calculation to algebra reminds the author of Nobelist Hans Bethe's remark that he only felt he understood a physical phenomenon when he could reduce it to algebra. This was quite evident when the author collaborated with Professor Bethe on a study of pion condensation in neutron stars some years ago.

[9] "One Hundred Years of the Eötvös Experiment", I. Bod, E. Fischbach, G. Marx and Maria Náray-Ziegler, August, 1990.

# 2

# Initial Probes of the Multiverse

Our universe is a 4-dimensional surface with complex-valued coordinates residing in a 16-dimensional space, also having complex-valued coordinates. We are "shielded" from seeing the complex-valued coordinates composing our universe because our instruments of measurement only measure real numbers. Happily, there is a group that we call the Reality group that transforms complex-valued time and space values to the real numbers we measure. This view of our universe directly leads to the known form of The Standard Model of Elementary Particles plus much more – including faster-than light motion. We have described this new theory in a series of books listed in the references.

The theory of Quantum Gravity developed by DeWitt and others culminating in the Wheeler-DeWitt equation, when generalized to complex space-time coordinates, leads to the cosmological features of our universe: an origin for the Cosmological Constant, an origin for the spatial asymmetry of the Universe, and an understanding of the origin for the newly found Web of Galaxies (that links all the groups of galaxies) in our universe. As part of this development we have previously shown that at least one sister universe must exist to play the role of an Observer, and to provide a "clock" for our universe in the Copenhagen formulation of quantum systems. Further we have shown that, if our universe is embedded in a flat Euclidean space, then the special role of inertial reference frames and the origin of inertia follow – thus eliminating speculations starting with Mach's book about the origin of these phenomena.

Working within this framework we have developed a theory of the multiverse consisting a 16-dimensional Flatverse (a flat, Euclidean space) with complex-valued coordinates within which reside universes such as our own.

In volume I of this book we described methods of travel via faster-than-light starships outside our universe to other universes in the multiverse. In this book we focus on the "slingshot mechanism" of volume I – the best hope – to escape our universe and travel in the multiverse in the distant future when the needed technology is available. We also will look at ways to probe the multiverse (this chapter) since the multiverse is not far away but in fact all around every point in our universe. (Remember our universe is a surface in the multiverse and thus we need only edge off this surface to be in the multiverse.) Mathematically, every neighborhood of a point on a surface contains an infinite number of points of the space in which the surface is embedded (except perhaps for space filling surfaces – an issue not relevant in the present considerations.)

## *2.1 Small Scale Microscopic Escape from our Universe*

Having seen that each point of our 4-dimensional space-time is "infinitely close" to an infinity of points of the Flatverse part of the multiverse, we now turn to inquire how we can infinitesimally escape into the Flatverse, at least in part. We can visualize this process in a three dimensional example: suppose the x-y plane is "our" universe; we propose to escape in the z-direction. See Fig. 2.1.

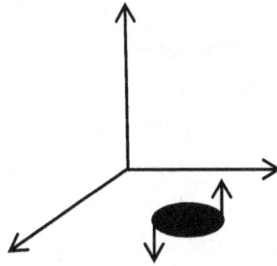

**Figure 2.1. An object that was flat on the x-y plane (z = 0) is twisted to be partly in the x-y plane (universe) and (microscopically) partly out of the x-y universe into the full 3-dimensional "universe."**

In the next section we will describe a possible mechanism for partly escaping from our universe so that part of an object is in the Flatverse while part of the object may still be in our universe – rather like the case depicted in Fig. 2.1.

In the corresponding case of our universe and the Flatverse we would see the part of the object still within our universe but we would not see the part of the object outside our universe since light emitted from the Flatverse part of the object would have to travel an infinite distance in the coordinates of our universe to reach us in our universe. Thus there is a type of horizon between our universe and the Flatverse with some slight resemblance to black hole horizons.

## 2.2 A Mechanism for Generating a Baryonic Field

In volume I we discussed several uniship (starship) designs for escaping from our universe into the multiverse. Later we will discuss the slingshot mechanism which we view as the preferred method for entry into the multiverse.

In this section we will consider a mechanism for short distance probes of the multiverse. We see its purpose as scientific research on the horizon between our universe and the multiverse as well as a preliminary to the eventual conquest of the multiverse tens of thousands of years from now (although there is a possibility of multiverse exploration within a few thousand years given key technology breakthroughs.)

The mechanism that we will describe is a device for the stimulated emission of baryonic radiation which will drive a probe completely into the multiverse or into a state where the probe is partly in our universe and partly in the multiverse. This partial entry is of particular interest as it can be used to sense multiverse properties and use the in-universe part for data acquisition and processing. If the probe can be completely embedded in the multiverse outside our universe, then interesting questions arise such as the reentry process into our universe and propulsion within the multiverse.

We shall use the acronym PHASER for a device that produces the stimulated emission of baryonic radiation since baryonic radiation like electromagnetic radiation is quantized by particles we called *plancktons* in volume I (the baryonic analogue of photons). PHASER represents Planckton Hyper Amplification by Stimulated Emission of Radiation. Like lasers, PHASERs produce coherent radiation. As we will see, our PHASER device would also produce laser radiation as a byproduct of the generation of coherent baryonic radiation.

A PHASER is similar to a free electron laser. However, it uses a beam of perhaps lead[10] ions which, individually, have a large non-zero baryon number instead of the electron beam used in a free electron laser. As Fig. 2.2 shows the ions are first accelerated to high energies – perhaps as high as, or higher than, the energy of ion beams created at CERN's Large Hadron Collider (LHC). They are then sent through a wiggler that causes the ions to oscillate perpendicular to the beam direction.   Due to the much larger mass of the ions compared to electrons

---

[10] It could use ions of other heavy metals such as gold or uranium (U-238) as the beam particle.

much more powerful magnet arrays must be used in the ion wiggler (undulator) to cause the ion oscillations that generate electromagnetic and also baryonic waves. The larger mass of the beam ions also reduces the amount of radiation generated by the oscillations. Thus the PHASER has two obstacles in the generation of electromagnetic and baryonic radiation. These can be overcome by using more powerful wiggler magnets and a higher energy ion beam.

In the wiggler phase an external laser of very high energy (shown in Fig. 2.2 as the bunching laser) can interact with the oscillating ion beam to cause bunching of the ions with the bunches separated by the optical wavelength of the external laser along the initial beam direction. This bunching causes the bunched ions to radiate electromagnetic radiation in phase (coherent radiation). It also causes the bunched ions to radiate baryonic radiation in phase with the same wavelength as the electromagnetic radiation giving coherent baryonic radiation.

Each type of radiation is thus enhanced as bunching gets more pronounced until it reaches a saturated power level that is orders of magnitude greater than the corresponding wiggler radiation levels. The wavelengths of the electromagnetic and baryonic radiations can be adjusted by manipulating the initial ion beam energy and the strength of the wiggler magnetic fields.

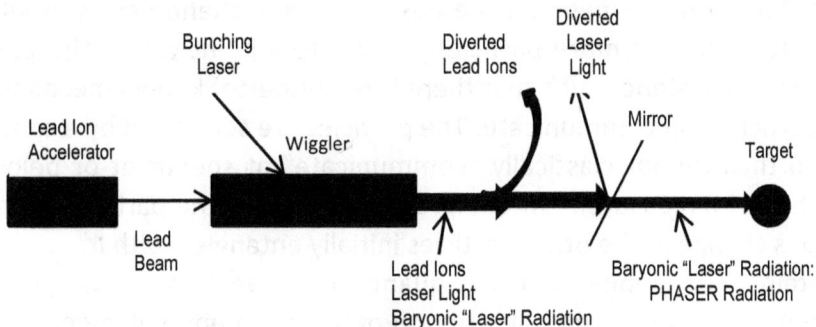

**Figure 2.2. A schematic diagram of a PHASER. – stimulated coherent Planckton radiation.**

Given the differing coupling constants of electromagnetic and baryonic radiation the relative power of the baryonic and electromagnetic coherent radiation should vary as a power of $\alpha_B/\alpha_B \simeq 5.9 \times 10^{-39}/7 \times 10^{-3} \simeq 10^{-36}$. Thus the coherent electromagnetic power is very much greater than the coherent baryonic radiation. As Fig. 2.2 shows the lead ion beam is diverted and the laser beam is also deflected from the output stream so that only the coherent baryonic radiation impinges on the target.

The baryonic radiation exerts pressure on the target accelerating it into an overlap region between our universe and the Flatverse since, as volume 1 shows, the magnetic baryonic field part of the baryonic radiation causes acceleration into dimensions of the Flatverse external to our universe's dimensions.

*Thus we have a device for entering the multiverse microscopically that can be developed today with (perhaps) more powerful magnets and lasers using (perhaps) the CERN LHC as the accelerator.* This conclusion depends on the existence and the value of the coupling constant of the baryonic field. Section 1.4 points out the need for decisive experiments on these issues.

## *2.3 Quantum Entanglement and the Multiverse*

In 1935 Einstein, Podolsky, and Rosen,[11] and shortly afterwards Erwin Schrödinger, began the discussion of what has become known as *quantum entanglement.* It has become a subject of growing, widespread interest since its validity has been demonstrated in numerous experiments.

Quantum entanglement can be described as a phenomenon in which a quantum state of two or more particles evolves to a point where the particles are separated by a distance such that there is no (hitherto) known mechanism by which the particles can communicate. The particles are separated by a space-like distance and thus cannot classically "communicate" at speeds at or below the speed of light. Yet a measurement of a property of one of the particles causes an instantaneous change in the other particles initially entangled with it.

Einstein's unhappiness with quantum entanglement is succinctly expressed with his often quoted description of quantum entanglement, "spooky action at a distance." Yet, since the 1930's, experiments have repeatedly shown that quantum entanglement is correct.

The bothersome issue which many physicists, starting with Einstein, are concerned is the mechanism by which instantaneous quantum effects happen.

---

[11] Einstein A, Podolsky B, and Rosen N, "Can Quantum-Mechanical Description of Physical Reality Be Considered Complete?", Phys. Rev. **47**, 777 (1935).

Dynamically we know that the evolution of a quantum system in the usual coordinates of our universe (whether a flat or curved space-time) yields quantum entanglement. However, a quantum measurement is outside the dynamical equations being a sharp transition between the quantum state before and after the measurement – only regulated by continuity conditions between the initial quantum state and the post-measurement quantum state.

In the case of a two particle system in an initial quantum state the invariant distance (squared) between the parts of the system when they are widely separated is

$$\Delta s^2 = \Delta t^2 - \mathbf{\Delta x}^2 \qquad (2.1)$$

with $\Delta t = 0$ at equal time. The coordinates are real-valued and governed by the transformations of the Lorentz group in the usual discussions of quantum entanglement. Thus the quandary of instantaneous changes in the parts of a system separated by a space-like distance.

We now suggest that quantum measurements should be viewed within the framework of complex-valued 4-dimensional space-time and the complex Lorentz group which we have shown in prior books leads to the form of The Standard Model of Elementary Particles (and thus quantum theory).[12]

If complex coordinates are the proper coordinates for observers then quantum entanglement is understandable. If a state is created and then parts of the state separate, measurement of part of the state at a certain time can set the other part of the state instantaneously. The invariant distance between the parts are space-like separated in real-valued space-time at equal time:

$$\Delta s^2 = - \mathbf{\Delta x}^2 = -(\Delta x^2 + \Delta y^2 + \Delta z^2) < 0 \qquad (2.2)$$

But they can be transformed by a complex Lorentz group transformation $R^i_j$

$$[\mathbf{\Delta x'}]^i = R^i_j [\mathbf{\Delta x}]^j \qquad (2.3)$$

to zero spatial distance giving

---

[12] We also showed that the Wheeler-DeWitt equation of quantum gravity when generalized to complex-valued General Relativity explains many of the features seen in the universe such as the web of clusters of galaxies and the lopsided nature of the universe. Thus we conclude that the multiverse and its universes have complex-valued coordinates. Some relevant references are Blaha (2011c), (2012a), (2012b), (2013a), (2013b), and (2014a). Much of the content of these books were introduced in earlier books by the author.

$$\Delta s'^2 = -\Delta \mathbf{x}'^2 = 0 \tag{2.4}$$

In this new coordinate system the time difference is not zero – it is a pure imaginary number $\Delta t' = i|\Delta t'|$ thus maintaining the invariance of the full invariant interval:

$$\Delta s^2 = \Delta t^2 - \Delta \mathbf{x}^2 = \Delta s'^2 = \Delta t'^2 - \Delta \mathbf{x}'^2 \tag{2.5}$$

Thus, using complex coordinates,[13] which are truly the fundamental coordinates of space-time, all spatial distances can be mapped to zero and the apparent separation suggesting quantum entanglement at a distance is a result of the map of complex-valued coordinates to the real-valued coordinates. The complex coordinates of our universe are related to the coordinates of the complex 16-dimensional Flatverse by

$$y_i = f_i(x) \tag{2.6}$$

for $i = 1, \dots, 16$ where x is a complex 4-vector in our universe and the $y_i$ are the corresponding coordinates in the Flatverse. The Quantum observer can make a measurement in either set of coordinates in a reference frame where the spatial location of the apparently separated parts of the quantum state coincide.

Spatial separations at equal time can thus be mapped to zero for observations (measurements) in complex coordinates while the dynamical evolution of the quantum state is expressed in real-valued coordinates. These real-valued coordinates can be related to the complex coordinates of our universe or the Flatverse.

*We conclude that there is no mysterious "spooky action at a distance" in quantum entanglement. The mystery is explained if we use the right coordinate system for the observer: complex-valued coordinates of the multiverse.*

*Thus all parts of a quantum state can be viewed as being located at the same spatial point in the right reference frame. Properly viewed, there is no mystery to quantum entanglement.*

*Returning to our subject: Despite the disappearance of mystery in quantum entanglement, it still remains an exciting topic in view of its potential*

---

[13] As shown in our earlier books complex-valued coordinates in any coordinate system can be mapped into the real-valued physical coordinates that we experience using the Reality group. See Blaha (2011c) and other books by the author for more details.

*applications such as communicating instantaneously at interstellar, intergalactic, and multiverse distances which we have discussed elsewhere.*

## 2.4 The Horizon between a Universe and the Flatverse

The horizon (boundary) between a complex 4-dimensional universe, and the complex 16-dimensional Flatverse in which it is embedded as a surface, has interesting properties. We will begin by considering the boundary of a simple universe. Then we will show that every point in our universe is "infinitely" close to points of the Flatverse.

### 2.4.1 Boundary of a Universe

The boundary (horizon) of a universe at a time $x^0$, assuming a non-pathological topology,[14] is the set of points defined by:

$$B = \{y \,|\, \forall y \text{ s/t } y = \lim_{|x| \to \infty} f(x^0, \mathbf{x})\} \tag{2.7}$$

In words, the boundary B at time $x^0$ is the set of all y values obtained from $f(x^0, \mathbf{x})$ (the 16-vector form of eq. 2.6) when the magnitude of the spatial vector argument $\mathbf{x}$ of $f(x^0, \mathbf{x})$ approaches infinity. Thus a universe is a 4-dimensional surface whose spatial coordinate values approach infinity as it approaches its boundary. However, the Flatverse coordinates of points on the boundary are finite.

A simple mathematical example illustrating the boundary of a universe is a 3-dimensional spherical universe embedded in a larger 3-dimensional universe (a Flatverse). We choose the center of the sphere and its coordinate system to coincide with the center of the multidimensional Flatverse universe coordinates. Using the mapping of eq. 2.6 in spherical coordinates the radial coordinate of this Flatverse can be related to the radial coordinate of the embedded 3-dimensional universe by

$$r_y = f(r) = a + 1/r \tag{2.8}$$

---

[14] Such as universes where going to an "end" of the universe merely positions one at a beginning point on the other side of the universe. If the universe is lopsided as recent data suggests then this particular topology seems unlikely.

As r $\rightarrow \infty$ we see $r_y \rightarrow$ a. Thus the embedded universe has infinite extent but lies within a finite region of the Flatverse in this example.

Thus the concept of a horizon of a universe in the Flatverse is understandable. A corollary of this concept and example is universes are open sets within the Flatverse. Their "edges" are not part of the universe. If we introduce quantum effects the "edges" have quantum "blurring."

### 2.4.2 Each Point in a Universe has Flatverse Points "Around" It

Another important fact is the "nearness" of Flatverse points to any point within our universe. This situation is easily visualized in a 3-dimensional Flatverse with the x-y plane being the universe. In the 3-dimensional neighborhood of any point in this universe there is an infinite number of points of the 3-dimensional Flatverse. Thus every point of the 2-dimensional universe is infinitely close to points of the 3-dimensional Flatverse. Every neighborhood of a point in the 2-dimensional universe contains an infinite number of points of the 3-dimensional Flatverse.

A similar situation applies to a 4-dimensional universe embedded in a 16-dimensional Flatverse. Every point of the universe is infinitely close to an infinite number of Flatverse points. Every neighborhood of a point in the 4-dimensional universe contains an infinite number of points of the 16-dimensional Flatverse.

### 2.4.3 Leakage from our Universe

If the neighborhood of every point in our universe has an infinite number of Flatverse points (outside our universe), then the question arises of "leakage" of matter and radiation out of the universe through classical or quantum modes. To great accuracy we know that leakage does not occur. If we inquire as to the cause of energy and matter conservation (lack of leakage) within the universe, we find the answer is implicit in the conclusion of subsection 2.4.1: although two points are *infinitesimally* far apart using the invariant distance measured in Flatverse coordinates, they are *infinitely* far apart if they are on opposite sides of a universe horizon. Thus any leakage would have to traverse an infinite distance and so it does not happen.

### 2.4.4 "Jumps" Out of Our Universe

The discussion of the preceding section shows that jumps out of our universe if mediated by processes strictly defined within our universe cannot happen. However if the jump process is mediated by a Flatverse process such as the baryonic force (defined in Flatverse coordinates) then the jump can happen as we show in our discussion of methods for uniships to escape the universe. *Thus the coordinates of a process can influence what it can and cannot do.*

### 2.4.5 Penetrating the Horizon to Escape into the Flatverse

Given the picture of the horizon presented above it is clearly easy to escape from the universe IF one can use the baryonic force to power the uniship. We describe this in volume I and will consider the neutron star slingshot mechanism in detail later. The uniship will enter the Flatverse and will initially be partly in and partly out of the universe. The uniship can then proceed to emerge totally out of the universe and travel in the Flatverse.

### 2.4.6 Reentering the Universe from the Flatverse

Given the baryonic force it is possible to totally escape from our universe. However reentry into our universe is problematic. First, the uniship has to be able to select the entry point based on a thorough knowledge of eq. 2.6 and its inverse to select the point of entry into our universe. This determination is not without issues since the inverse of eq. 2.6 is defined only up to a subspace of the Flatverse. The inverse of eq. 2.6 can be formally expressed as

$$x_\mu = f^{-1}{}_\mu(y_1, .. , y_{16}) \tag{2.8}$$

where $\mu = 0, ... , 3$. Eq. 2.8 expresses four equation for the x coordinates in sixteen unknowns – the Flatverse coordinates $y_i$. Thus there are 12 Flatverse degrees of freedom (dimensions) in the determination of each universe coordinates x. Given the functions $f_i(x)$ for $i = 1, ... , 16$ the y coordinates can be determined up to the 12 extra degrees of freedom permitting the determination of the entry point from the Flatverse to a specified point in the universe.

Assuming we have determined the correct entry point we have to confront the issue of the 16-dimensional nature of the uniship engine. Effectively it can be view as having 16 thrust ports – one pointing in each direction of the Flatverse. Upon entry into our universe these thrust ports, and their engine, will still be in the Flatverse unless a mechanism is used to fold the thrust ports, like an umbrella, into directions within our universe. We will discuss this issue in chapter 8 – the opening and closing of the thrust port umbrella.

If we are unable to construct a mechanism to close the umbrella then the returning uniship will resemble a shark whose fin sticks above the water while the body is below the water. In the uniship case the fin would stick into the Flatverse. This situation may not be a negative. Having the engine in the Flatverse may make the future use of the uniship for new travels easier.

### 2.4.7 Objects "Straddling" the Border between our Universe and the Flatverse

When a uniship or other extended object is exiting the universe at some velocity the question arises of the state of the object when it is partially in and partially out of the universe. We know that the object being 4-dimensional will continue to be 4-dimensional, barring effects of the baryonic force that would "twist" parts of the object into additional dimensions as discussed in chapter 4.

There are two simple reasons for this view: 1) if we take a string and move it from two dimensions to three dimensions, it retains its one dimensionality; 2) when an object such as an asteroid or planet passes through the horizon of a black hole it retains its dimensionality and topology.

Similarly a uniship transit from a universe to the Flatverse retains its dimensionality modulo baryonic force effects.

### 2.4.8 The Different Physical Implications of Flatverse Vs. Universe Cordinates

A change of coordinate system by a linear transformation such as a Lorentz transformation or a general coordinate transformation does not change the physical situation – there are clear transformation laws for physical quantities and equations that preserve the physics of a phenomenon.

However, the "transformation" law (eq. 2.6 and its "inverse" eq. 2.8) relating the coordinates of a universe to the coordinates of the Flatverse in which it resides does not necessarily maintain the physics of a phenomenon. Since the Flatverse is 16-dimensional and a universe is necessarily 4-dimensional according to the reasoning of Blaha (2011c), and more recent books, the inverse transformation (eq. 2.8) is not uniquely defined. This is especially clear in the case of a *flat universe* which may well be the case of our universe. If a universe is flat then eq. 2.6 is a linear transformation. In this case an "inverse" transformation (eq. 2.8) has a null space of 12 dimensions such that an element $y_n = (y_{n1}, .. , y_{n16})$ in the null space is mapped to zero:

$$x_\mu = f^{-1}{}_\mu(y_{n1}, .. , y_{n16}) = 0 \qquad (2.9)$$

The points in the remaining four dimensions of the Flatverse domain constitute the domain of the inverse transformation.

The physical consequence of this case, and the more general case of non-linear transformations of the form of eqs. 2.6 and 2.8, is that the escape from a universe, if performed dynamically in Flatverse coordinates, happens in an incrementally continuous fashion as described in section 2.1 and chapter 4. However physical processes defined strictly in terms of universe coordinates $x^\mu$ cannot proceed to escape from a universe incrementally since the universe is of infinite extent in its own coordinates. Thus an infinite motion would be required in universe coordinates to exit the universe into the Flatverse. In section 2.4.4 we showed this infinite motion requirement results in no leakage from our universe despite the fact that all neighborhoods of any point in a universe contain an infinite number of Flatverse points. "So near in Flatverse coordinates but yet so far in universe coordinates."

*The physics of Flatverse coordinates is different from the physics of universe coordinates.*

## 2.5 An Improbable Multiverse Connection to Spirits, the "Spirit World", and UFOs

Ghosts, spirits, and spiritual phenomena have been associated with "the fourth dimension" and other extra dimensions since the 1850s. UFOs are also often described as coming from or through other dimensions. Some scientists have also associated physical phenomena with other dimensions.

We do not believe these characterizations of phenomena as artifacts of other dimensions to be correct in the manner in which they are described.

However, we do think that one can simulate spiritual-like phenomena and UFO-like phenomena using the multiverse. This section will briefly outline these possibilities. The hope is to forestall attempts to use our multiverse theory to bolster support for these phenomena.

### 2.5.1 "Spiritual" Phenomena and the Multiverse

Spiritual phenomena have several types: the appearance of visions of people or things of one sort or another; material objects passing through solid objects; unseen voices; and so on. All of these phenomena could be simulated using the multiverse to make things appear from the multiverse or to go from within the universe through the multiverse and reappear in our universe again.

Some examples on multiverse simulations are:

1. Using a "Sidestep" into the Multiverse to Circumvent Solid Obstacles in our Universe.

2. Going through a Solid One Dimensional Wall.  A 1-dimensional example is
   ------------------|----------------

3. Persons or things going through Solid Three Dimensional Walls.

While these "partly in and partly out of the universe" phenomena can be simulated by a multiverse agent, the agent's nature and purpose are not determinable from physical considerations of the multiverse.

## 2.5.2 UFO Phenomena and the Multiverse

UFOs have been "seen" in the skies in many regions of the earth. They are often characterized as having high speeds, extremely high accelerations, and the ability to make abrupt changes in direction at high speeds. All of these phenomena are understandable if we are seeing objects in the multiverse where modest changes in speed, acceleration and direction in multiverse coordinates can map to the UFO movements that we see in our universe's coordinates.

Beyond stating these phenomena can be understood from a multiverse perspective we can say no more. They may be real. Their purpose, if they are real, cannot be discerned. Multiverse physics cannot enlighten us on these subjects.

# 3

# Voyages into the Multiverse

Traveling in in the multiverse between universes places extraordinary demands on uniships. Speed, acceleration, and fuel requirements are the primary issues because the distance between universes is vast and human lifetimes are short. We should like to be able to travel fairly quickly between universes. If one wants colonization, commerce, and timely exploration then the trip between two universes, on average, should be perhaps six months to a year of earth time. (Time on a uniship proceeds much, much more rapidly than earth time. But we can circumvent this potential problem using suspended animation for passengers and very long-live machinery for the uniship and its contents.)

In this chapter we will consider Uniship distance and speed related requirements for "short" distance travel in the multiverse.

## 3.1 The Distance Scales of Our Universe and the Multiverse

We will begin with a consideration of distance scales between galaxies in our universe and anticipated distance scales between universes in the multiverse. Our universe has a web of groups of galaxies. Typically the distances between galaxies in a group are of the order of several million light years.[15]

The distances between universes in the multiverse are a matter of conjecture. However if we take the order of magnitude of the ratio of the size of a galaxy (say the Milky Way which is 100,000 light years in diameter) to the separation of galaxies in a group of galaxies (say three mllion light years) as a guide (a factor of 30) and use the same ratio with the size of our universe as the input, then, with the size of the (visible) universe being about 50 billion light years the order of magnitude of the relative separation between universes could be roughly estimated to be perhaps two trillion light years.

---

[15] BOSS – Baryon Oscillation Spectroscopic Survey and other studies of WMAP data.

This estimate is at best an order of magnitude estimate. It suggests that multiverse universe distances are of the order of 1,000 times distance scales in our universe – a not unreasonable value.

## 3.2 Starship Distance and Speed Requirements in Our Universe

In our book *All the Universe!* (and in earlier books) we developed the theory of faster-than-light starships for travel between stars and galaxies in our universe. For the reader's convenience we reproduce part of *All the Universe!* It appears that speeds of 60,000c give acceptable travel times (up to a year) within our galaxy, and *speeds of a few million c give acceptable travel times to nearby galaxies where c denotes the speed of light.*

At 3,000,000,000c a universe that is 2,000,000,000 light years away could be reached in eight months (neglecting acceleration and deceleration times) – an acceptable time for trade, exploration and possibly colonization. The fundamental problem is to develop an energy source that can fuel such enormous speeds – a reason for anticipating 50,000 years of necessary development time.

The basic dynamical considerations for faster-than-light travel that are described in our earlier books are:[16]

### 3.2 Superluminal Starship Dynamics

In this section we will consider a constant, propulsive force in a starship's rest frame that drives the starship from a sub-light velocity to a superluminal velocity. The key factor in achieving a superluminal speed is evading the singularity in $\gamma$ at $v/c = 1$. We accomplish this goal by having a complex force – a force with a real and imaginary part – that generates a complex acceleration, and thus a complex velocity, that "goes around" the singularity in $\gamma$ in the complex velocity plane. We assume that an "instantaneous" Lorentz transformation relates the earth reference frame and the starship reference frame.

We assume a constant, complex force exists in the rest frame of the starship due to the starship's thrust in the direction of the positive x' (and x) axis. The starship (primed coordinates) and earth (unprimed coordinates) coordinates have parallel axes as in Fig. 2.1. The spatial force in the positive x direction is

$$\mathbf{F'} = g\hat{\mathbf{x}} \tag{3.1}$$

---

[16] This excerpt is from Blaha (2011b). It is *revised* and shortened. We have modified the presentation to include comments on *physical* speeds and distances. These quantities are the absolute values of the complex quantities. They are obtained by applying a Reality group transformation to the complex v and x values. The Reality group, which leads to The Standard Model of Elementary Particles, is described in Blaha (2011c), (2012a), (2013b) and other books by the author.

where g is a complex constant.

The fourth component of the force (since force is a Lorentz 4-vector) is zero in the starship's rest frame:

$$F'^0 = 0 \tag{3.2}$$

Applying the inverse of the Lorentz transformation between earth and the starship coordinates we find the force in the earth rest frame is

$$F^0 = \gamma(F'^0 + \beta F'^x/c) = \gamma\beta F'^x/c = \gamma vg/c^2 \tag{3.3}$$
$$F^x = \gamma(F'^x + \beta cF'^0) = \gamma F'^x = \gamma g$$
$$F^y = F^z = 0$$

where $\beta = v/c$, c is the speed of light, and $\gamma = (1 - \beta^2)^{-\frac{1}{2}}$ as before. We again use the superscripts x, y, and z to identify the components of the spatial force. The spatial momentum of an object of mass m is

$$\mathbf{p} = \gamma m\mathbf{v} \tag{3.4}$$

and the dynamical equation of motion is

$$d\mathbf{p}/dt = \mathbf{F} \tag{3.5}$$

in the "earth" coordinate system resulting in

$$dp^x/dt = \gamma g \tag{3.6}$$

with[17]

$$dp^y/dt = dp^z/dt = 0 \tag{3.7}$$

The differential equation resulting from eq. 3.5 is

$$d(\gamma v)/dt = \gamma g/m \tag{3.8}$$

Assuming initially that g is real we must use $\gamma = (1 - \beta^2)^{-\frac{1}{2}}$ for v < c and $\gamma = (\beta^2 - 1)^{-\frac{1}{2}}$ for v > c based on the need for real coordinates for faster than light travel as expressed in eqs. 2.1 and 2.2. The solutions for real v are[18]

v < c, Re $v_0$ < c

$$v = c\{1 - 2/(1 + ((c + v_0)/(c - v_0))\exp[2g(t - t_0)/(mc)])\} \tag{3.9a}$$

---

[17] There is thrust in the y and z direction as well. To avoid getting distracted by the details of an exact calculation we approximate the force in those directions as zero.

[18] The velocity is entirely in the x-direction in this calculation. It can, and does, have complex values in this example. See footnote 45 in the discussion of our starship engine to see how the complexity of the value arises.

<u>$v > c$, Re $\acute{u}_0 \geqslant c$</u>

$$v = c\{1 - 2/(1 + ((c + \acute{u}_0)/(c - \acute{u}_0))\exp[2\breve{g}(t - \mathfrak{t}_0)/(mc)])\}\qquad(3.9b)$$

where the velocity is $v_0$ at time $t_0$, $\acute{u}_0$ is the velocity[19] at $t = \mathfrak{t}_0$ and $\breve{g}$ is the acceleration for Re $v \geqslant c$.[20]

Analytically continuing eqns. 3.9 to complex $v$ with a complex force constant $g$ we obtain the starship equation of motion. We require continuity when the real part of $v = c$ by requiring that when Re $v(t)$ of eq. 3.9a equal $c$, that $\mathfrak{t}_0 = t$ and $v(\mathfrak{t}_0)$ of eq. 3.9a equal $\acute{u}_0$. These conditions fix $\mathfrak{t}_0$ and $\acute{u}_0$.

<u>Note:</u>

Eqs. 3.9 can easily be integrated to give the distance traveled in the x direction.

<u>$v < c$, Re $v_0 < c$</u>

$$x = x_0 + (mc^2/g)\ln((1 - v_0/c + (1 + v_0/c)\exp[2g(t - t_0)/(mc)])/2) - c(t - t_0)\quad(3.10a)$$

<u>$v \geqslant c$, Re $\acute{u}_0 \geqslant c$</u>

$$x = x_0 - (mc^2/\breve{g})\ln((1 - \acute{u}_0/c + (1 + \acute{u}_0/c)\exp[2\breve{g}(t - \mathfrak{t}_0)/(mc)])/2) - c(t - \mathfrak{t}_0)\quad(3.10b)$$

or, correspondingly,

<u>$v < c$, Re $v_0 < c$</u>

$$x = x_0 + (mc^2/g)\ln[(1 - v_0/c)/(1 - v/c)] - c(t - t_0)\qquad(3.11a)$$

<u>$v \geqslant c$, Re $\acute{u}_0 \geqslant c$</u>

$$x = x_0 + (mc^2/\breve{g})\ln[(1 - \acute{u}_0/c)/(1 - v/c)] - c(t - \mathfrak{t}_0)\qquad(3.11b)$$

The complexity of $g$ and thus of the velocity causes $x$ to be complex. The starship is thus generally at a point $x$ in complex space. Its actual physical location using the Reality group of Blaha (2011c) to transform the complex coordinates to real-valued coordinates is the absolute values of $x$, $|x|$.

Superluminal travel to a distant star (or galaxy eventually) requires three phases in general. In the first phase the starship accelerates with a value for the thrust $g$ that enables it to reach a high complex velocity whose real part was much greater than the speed of light. In the second phase the starship coasts to a point "not far" from the destination. At this point the starship is located at a point in complex space which has an equivalent real-valued, physical, spatial location. In the third phase the starship engines are turned on and the thrust decelerates to bring the starship to its destination (located at a real-valued coordinate position.)

---

[19] It is greater than c by assumption in the calculation of eq. 3.9b.

[20] Although eqs. 3.9a and 3.9b have the same form, the acceleration for Re $v < c$ can be changed to a new value $\breve{g}$ after Re $v$ exceeds the speed of light in order to approach the singularity discussed later.

### 3.3 Achieving High Superluminal Starship Velocities

To achieve faster-than-light motion the constant force value $\breve{g}$ required after Re $v \geqslant c$ must satisfy a special set of conditions. These conditions emerge from a consideration of the denominator of eq. 3.9b:

$$1 + ((c + \acute{v}_0)/(c - \acute{v}_0))\exp[2\breve{g}(t - t_0)/(mc)] \qquad (3.12)$$

where $\acute{v}_0 \geqslant c$. If this denominator approaches zero then the speed v becomes infinite if g has an appropriate complex value. Let

$$\breve{g} = g_1 + ig_2 \qquad (3.13)$$

If we wish the velocity to get very large (approach infinity) after some acceleration time interval $\triangle t = t_1 - t_0$ we set

$$1 + ((c + \acute{v}_0)/(c - \acute{v}_0))\exp[2\breve{g}\triangle t/(mc)] = 0 \qquad (3.14)$$

with the result infinite velocity occurs when

$$g_2 = (mc/(2\triangle t))\{n\pi + Im\ ln[(c - \acute{v}_0)/(c + \acute{v}_0)]\} \qquad (3.15)$$

and

$$g_1 = (mc/(2\triangle t))\ Re\ ln[(c - \acute{v}_0)/(c + \acute{v}_0)] \qquad (3.16)$$

where n is an odd, positive integer, since $\acute{v}_0$ is complex in general. Eqs. 3.15 and 3.16 enable the real part of the velocity, and thus the physical velocity, to become infinite in the time interval $\triangle t$. We assume n = 1 in the following discussions. Substituting in eq. 3.9b we obtain the complex-valued velocity v. The instantaneous physical velocity is $|v|$.

$$v = c\{1 - 2/[1 + ((c + \acute{v}_0)/(c - \acute{v}_0))^{1 - (t - t_0)/\triangle t}\ e^{in\pi(t - t_0)/\triangle t}]\} \qquad (3.17)$$

### 3.4 Example of a Starship Accelerating from Sublight speed to the Speed of Light

We will now consider a specific example of the acceleration phase of a starship to get a feeling for an *optimal* "low" speed case.[21] First we note that the acceleration of a rocket of mass m with a propellant exhaust speed $v_e$ in the rocket's rest frame is given by

$$dv'/dt' = (v_e/m)\ dm/dt' \qquad (3.18)$$

and thus the constant g of eq. 3.1 is[22]

---

[21] The value of the imaginary speed is quite high. In realistic cases such as those of the first generation of starships the value of the imaginary speed will undoubtedly be much lower – probably of the order of magnitude of the speed of light. Thus first generation starships will travel much more slowly but still much faster than the speed of light.

[22] In the illustrative example we will consider the acceleration g changes to $\breve{g}$ when Re v = c. The acceleration in a real world case would vary in such a way as to maximize speed while minimizing energy consumption.

$$g = mdv'/dt' = v_e \, dm/dt' \qquad (3.19a)$$

Since we intend to generate the thrust with a quark-gluon plasma producing an extremely high-energy exhaust we will *choose* the value of the starship acceleration to be equal to the acceleration due to gravity at the earth's surface times $(1 + i)$:

$$g/m = (1 + i)g_E = (1 + i)980 \text{ cm/sec}^2 \qquad (3.19b)$$

where m is the mass of the starship. If we further choose the thrust exhaust velocity $v_e$

$$v_e = -c - ic \qquad (3.20)$$

which is a reasonable choice for the exit speed thrust of the fireball then

$$dm/dt' = -3.27 \times 10^{-8} \text{ m} \qquad (3.21a)$$

in our example. If we envision a starship of 10,000 metric tons[23] then

$$dm/dt' = -327 \text{ gm/sec} \qquad (3.21b)$$

To reach light speed from $v = 0$, we note that the acceleration (eq. 3.19b) yields equal real and imaginary velocities so that $\gamma = 1$ and thus eq 3.8 simplifies to

$$dv/dt = g/m \qquad (3.22)$$

with the solution: $v = gt/m$. Substituting eq. 3.19b we find v reaches $c + ic$ in approximately 2 years. If we accept an acceleration at twice earth gravity the travel time is about one year of earth, *and starship,* time. Since the starship occupants will be in suspended animation a higher acceleration is acceptable. The amount of fuel expended in this example is nominally 10,171 metric tons – more than the mass of the starship. However, this value can be reduced significantly by doubling or tripling[24] the exhaust velocity $v_e$. Also there can be a significant transformation of energy into matter in the quark-gluon fireball that would also reduce the amount of stored fuel that was expelled.

### 3.5 From Light Speed to Enormous Speeds

In the example given in section 3.4 we considered an illustrative example of a starship accelerating to light speed. In this section we consider the second part of the acceleration: from light speed to enormous speeds taking advantage of the mechanism described in section 3.3. We will use an approximation to eq. 3.9b as its denominator approaches zero. Letting $t = t_1 + \tau$ where $\tau$ is small, and letting $\Delta t = t_1 - t_0$ then eq. 3.9b becomes

---

[23] About one-fifth the mass of the ship Queen Elizabeth.
[24] An expelled mass decrease by a factor of ½ or 1/3.

$$v = c\{1 - 2/(1 + ((c + \acute{u}_0)/(c - \acute{u}_0))\exp[2\breve{g}(\triangle t + \tau)/(mc)])\}$$
$$= c\{1 - 2/(1 - \exp[2\breve{g}\tau/(mc)])\}$$
$$\simeq c\{1 - 2/(1 - (1 + 2\breve{g}\tau/(mc))\}$$
$$\simeq c\{1 + (mc/\breve{g})(1/\tau)\}$$
$$\simeq (\breve{g}*mc^2/|\breve{g}|^2)(1/\tau) \tag{3.23a}$$

Thus the magnitude of the *physical* instantaneous velocity is

$$|v| \simeq mc^2/(|\breve{g}|^2|\tau|) \tag{3.23a'}$$

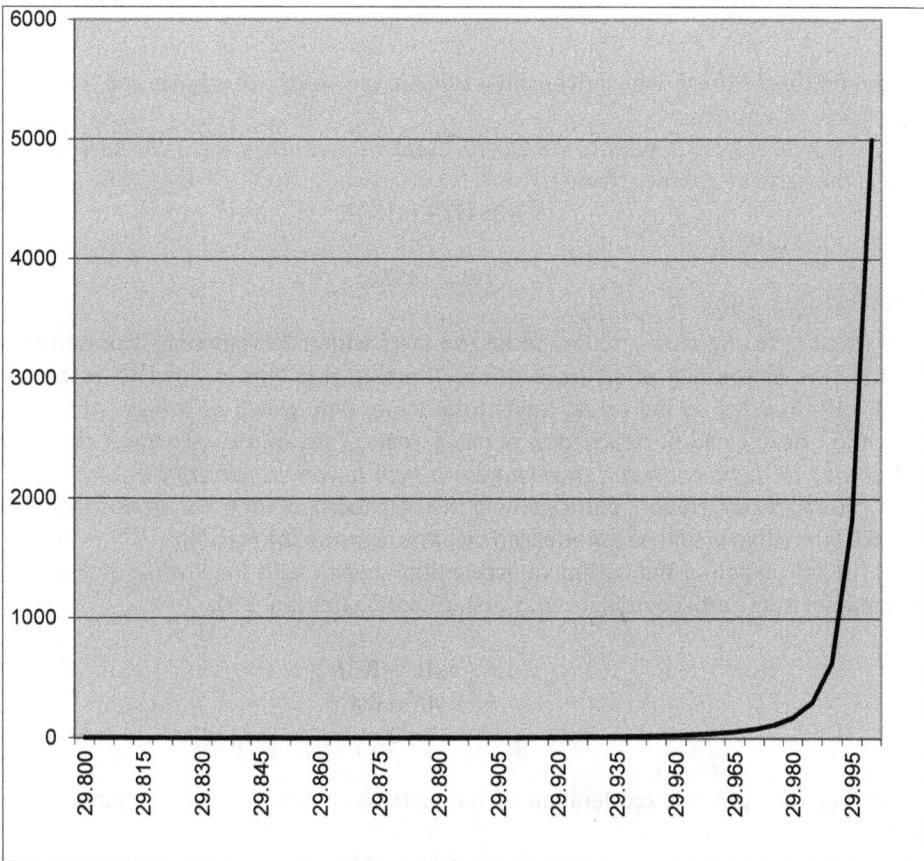

**Figure 3.1.** A plot of the real part of the velocity of a starship on its 29th and 30th earth day of travel up to 5,000c. The dynamics of this case are described in the text where the real speed reaches 8547c and beyond. Time is measured in earth days. Note: as the speed of the starship increases rapidly near the singularity point, time on the starship also passes more quickly so that the starship occupants do not experience very high acceleration. Starship time t' ≈ βt when β ≫ 1 where t is earth time.

Continuing the preceding illustrative example with $\acute{v}_0 = c + ic$, and m = 10,000 metric tons, and choosing $\triangle t$ = 30 days we find

$$\breve{g} = -4.66 \times 10^{13} + i6.43 \times 10^{13} \text{ gm-cm/sec}^2 \qquad (3.23b)$$

Given the signs of $g_1$ and $g_2$ we see that

- For small negative $\tau$ both the real and imaginary parts of v approach $+\infty$ as $\tau \to 0$ from below.
- For small positive $\tau$ both the real and imaginary parts of v approach $-\infty$ as $\tau \to 0$ from above.

At point in time close to the singularity point a starship can switch off engines and "coast" at high speed towards its destination.

At t = 30 − $1 \times 10^{-13}$ days ($\tau$ = 8.64×$10^{-9}$ sec) we find using eq. 3.23a that the starship velocity in the earth's reference frame is

$$v = 8547c + i11802c \qquad (3.23b)$$

and the physical velocity is

$$|v| = 14572c \qquad (3.23b')$$

At 14,572c any of the 100 or so known stars within 21 light years can be reached in about 1.5 days of coasting after acceleration. There is also time needed to accelerate and decelerate the starship so the actual travel total travel time would be longer. At 30,396c any point in the galaxy could be reached in about 4 years. *Thus Milky Way travel times become comparable to 16th century oceanic travel times via ships to various parts of the world!*

The rapid acceleration, particularly in the neighborhood of $\tau$ = 0 raises the question of the inertial forces that would be experienced by passengers on the starship.

The calculation of the maximum acceleration begins with the inverse of the relativistic transformation from earth coordinates to starship coordinates (eq. 3.3):

$$F^{\prime 0} = \gamma(F^0 - \beta F^x/c)$$
$$F^{\prime x} = \gamma(F^x - \beta c F^0) \qquad (3.24)$$
$$F^{\prime y} = F^{\prime z} = 0$$

which implies the apparent acceleration of the starship calculated in the starship's reference frame is

$$a' = F^{\prime x}/m = \breve{g}/m \qquad (3.25)$$

whereas in the earth's reference frame, the acceleration a is given by the derivative of eq. 3.9b.

$a = dv/dt$
$= 4(\breve{g}/m)((c + \acute{v}_0)/(c - \acute{v}_0))\exp[2\breve{g}(t - \acute{t}_0)/(mc)]/\{1 + ((c + \acute{v}_0)/(c - \acute{v}_0))\exp[2\breve{g}(t - \acute{t}_0)/(mc)]\}^2$
$$\qquad (3.26)$$

At $t = t_1 + \tau$ we see

$$a \simeq -mc^2/(\breve{g}\tau^2)[1 + 2\breve{g}\tau/mc] \approx -mc^2/(\breve{g}\tau^2) \tag{3.26a}$$

in the earth's reference frame.

### 3.6 The Acceleration Experienced on the Starship

The acceleration experienced by the starship occupants, the relevant acceleration for the occupants, is

$$a' = \breve{g}/m = -4.66 \times 10^3 + i6.43 \times 10^3 \text{ gm-cm/sec}^2$$
$$= (-4.75 + i6.75)g_E \tag{3.27}$$

by eq. 3.25 and 3.23b. This acceleration (experienced by the starship occupants) would be too high if they were not in suspended animation. In suspended animation they might be able to survive in somewhat high accelerations of the magnitude of eq. 3.27.

*If not, then the one month acceleration time of this example must be made somewhat longer. A longer acceleration time of, for example, 5 months[25] would still be acceptable for human travel in months or years to stars in our celestial neighborhood.* It is also possible that a form of robotic ship could use high accelerations, such as the above, to explore the far reaches of the galaxy.

### 3.7 Travel Time on the Starship – Suspended Animation

Another issue is the travel time experienced by the starship occupants: it is much, much longer than that measured on earth for high speeds. For example if v = 5,000c then starship time intervals will be approximately 5000 times longer than earth time intervals. A 2 month trip from earth's view would take around 1,000 years from the view of the occupants of the starship. *Therefore a practical method of suspended animation must be found for long distance travel. A 4 month round trip to a star would require the starship occupants to be in suspended animation for approximately 1,700 years – starship time. With suspended animation they could be kept biologically roughly "in sync" with the earth measured travel time of 4 months (plus time spent at the destination) despite the starship elapsed time of about 1,700+ years.*

### 3.8 Constant Superluminal Starship Travel

Having reached an enormous *real* speed such as a speed between 5000c and 30,000c we can turn off superluminal engines. The starship then moves at this constant speed in the absence of forces (and neglecting gravity and other minor perturbative forces). At a real speed of 5000c any place in the galaxy is a short travel time away. And nearby galaxies are reachable as well. Figure 3.2 shows the time required to reach various interesting destinations at 30,000c.

---

[25] Increasing $\triangle t$ by a factor of 5 would reduce a' by a factor of five.

| Destination | Distance (ly) | Approximate Travel Time (years) |
|---|---|---|
| To the other end of the Milky Way Galaxy | 100,000 | 3 |
| To the Center of the Milky Way | 30,000 | 1 |
| Large Magellenic Galaxy | 150,000 | 5 |
| Small Magellenic Galaxy | 200,000 | 7 |
| Andromeda Galaxy | 2,000,000 | 70 |

**Figure 3.2. "Coasting" part of travel time to various destinations at a real velocity of 30,000c.**

*Since much, much higher "coasting" velocities are also possible almost the entire visible universe becomes accessible to Mankind.*

## 3.3 Uniship Distance and Speed Requirements for "Short" Distances in the Multiverse

In this section we will consider some issues associated with uniship "short distance" trips in the multiverse such as a three trillion light year trip to a nearby universe.[26] The primary issue is the energy (and fuel) required to make a trip at high speed so that the travel time is of the order of months. Another important issue is the acceleration times that are required to attain high speeds.

### 3.3.1 Energy Required to Attain High Velocities

The energy E required to reach a high velocity much greater than the speed of light is not a simple mathematical expression in the speed v because the energy of an object moving much faster than the speed of light approaches zero. (The momentum approaches mc where m is the mass of the object. Thus $E^2 - c^2 p^2 = -m^2 c^4$ for $v > c$. The object is tachyonic.)

We will calculate the energy required to reach a speed $v > c$ in the earth's rest frame using the development of section 3.2 of this book. The energy E expended in the spatial x interval from $x_0$ to x is defined as

$$E = \int_{x_0}^{x} F dx = \int_{t_0}^{t} F v dt \tag{3.1}$$

---

[26] We will use the results of the previous section for acceleration and velocity in the x direction. The general case follows directly from this special case.

where $t_0$ is the initial time and t is the final time. Substituting from eq. 3.3 of the previous section we obtain

$$E = \int_{t_0}^{t} Fvdt = \int_{t_0}^{t} g\gamma vdt \qquad (3.2)$$

Eq. 3.2 can be transformed to the form:

$$E = (m/2) \int_{v_0}^{v} dv\, \gamma^{-1}\, d[(\gamma v)^2]/dv \qquad (3.3)$$

The exact form of the integral is:

$$E = [m\gamma(v^2 + c^2)/2 + mc^2/(2\gamma)] - [m\gamma_0(v_0^2 + c^2)/2 + mc^2/(2\gamma_0)] \qquad (3.4)$$

where $\gamma_0 = (1 - v_0^2/c^2)^{-1/2}$. Since the velocity is in general complex-valued, v and γ are also complex-valued as is E.

Using the Reality group that we introduced in previous books, the physical velocity is the absolute value of v, |v|, and the physical value of E is |E|. |E| is calculated by first substituting the complex values of v, γ, $v_0$ and $\gamma_0$ in eq. 3.2 and then taking the absolute value of the resulting complex quantity E.

We begin by defining

$$E_{part}(v) = m\gamma(v^2 + c^2)/2 + mc^2/(2\gamma) \qquad (3.5)$$

The integral in eq. 3.2, although superficially real-valued has complex quantities in the integrand. The result of the integration from a speed below the speed of light to a speed greater than the speed of light can be written as

$$E_{tot} = E_{part}(c - i\varepsilon) - E_{part}(v_0) + E_{part}(v) - E_{part}(c + i\varepsilon) \qquad (3.6)$$

as $\varepsilon \to 0$ due to the singularity at v = c. Since we are interested in the energy required to reach ultra-high speeds of the order of tens of thousands to trillions of times the speed of light, we see that

$$E_{tot} \to \lim_{v \to \infty} E_{part}(v) = -imcv/2 \qquad (3.7)$$

to leading order in v.  Thus

$$|E_{tot}| \rightarrow mc|v|/2 \qquad (3.8)$$

or, expressed in a more convenient way:

$$|E_{tot}| \rightarrow \tfrac{1}{2} E_{rest}|v|/c \qquad (3.9)$$

as $v \rightarrow \infty$ in the earth's reference frame where $E_{rest} = mc^2$ is the rest mass-energy of the entire ship including the fuel. Knowing the desired cruising speed v of a starship or uniship the ship must have an energy supply equal to $2|E_{tot}| = mc|v|$ when it leaves the earth's reference frame to provide for acceleration *and* *deceleration* plus the additional energy needed for other ship functions. For a round trip an energy supply of $4|E_{tot}| = 2mc|v|$ would be needed for the ship's engines.

### 3.3.2 Acceleration Time Required to Reach Extremely High Speed

We can determine the acceleration time required to reach velocity v from eq. 3.9 of the previous section:

$$v = c\{1 - 2/(1 + ((c + v_0)/(c - v_0))\exp[2g(t - t_0)/(mc)])\} \qquad (3.9a)$$

Inverting eq. 3.9 above yields the acceleration time interval required to achieve a speed v in the earth's reference frame:

$$t(v) - t_0 = (mc/2g)\ln\{[(c - v_0)/(c + v_0)]\,[(c + v)/(c - v)]\} \qquad (3.10)$$

where $v_0$ is the speed at $t_0$. For large v in the earth reference frame the time interval approaches

$$|t(v) - t_0| \rightarrow (mc/2g)\ln[(c - v_0)/(c + v_0)] + mc^2/(gv) \qquad (3.11)$$

to leading order in v. As $v \rightarrow \infty$ the interval becomes a constant:

$$|t(v) - t_0| \rightarrow (mc/2g)\ln[(c - v_0)/(c + v_0)] \qquad (3.12)$$

The reason for this limit on the earth reference frame time interval can be understood when one realizes that the corresponding ship time interval approaches infinity. Ship time increases much faster at high speeds greater than √2c. We shall see this in the next subsection.

### 3.3.3 Speed, Distance and Acceleration Time in the Starship/Uniship Reference Frame

In this subsection we will calculate dynamical quantities from the point of view of the starship/uniship reference frame.

In case examined in the previous section we found the acceleration in the earth reference frame to be $\gamma g/m$ by eq. 3.3. In the ship reference frame the acceleration is $g/m$. See the quoted section 3.6 within the previous section for more detail.

The transformation law from the earth reference frame (unprimed coordinates) to the starship/uniship reference frame (primed coordinates) in which the ship is moving at instantaneous speed v in the x direction is

$$t' = \gamma(t - \beta x/c) \tag{3.13}$$
$$x' = \gamma(x - \beta ct)$$
$$y' = y$$
$$z' = z$$

Substituting

$$x = x_0 + (mc^2/g)\ln[(1 - v_0/c)/(1 - v/c)] - c(t - t_0) \tag{3.11a}$$

from the previous section and

$$t(v) - t_0 = (mc/2g)\ln\{[(c - v_0)/(c + v_0)]\ [(c + v)/(c - v)]\} \tag{3.10}$$

from this section with $t_0 = 0$ and $x_0 = 0$, we obtain the ship distance and time to be

$$t' = \gamma\{(m(c - v)/2g)\ln\{[(c - v_0)/(c + v_0)]\ [(c + v)/(c - v)]\} - (mv/g)\ln[(c - v_0)/(c - v)]\} \tag{3.14}$$

Eq. 3.14 gives starship time as a function of speed. Thus the faster a ship speeds the more quickly time passes on the ship. As a result ships should have long lifetime equipment and place passengers in suspended animation for long periods of time.

For large v, the absolute value of the ship time approaches

$$|t'| \rightarrow |mc/g)\ln v| \qquad (3.15)$$

Thus the ship time approaches infinity as the ship speed approaches infinity. This situation explains the limit on earth time in eq. 3.12. It corresponds to ship time approaching infinity. (The ship speed approaches infinity in this limit as well.) Of course the limit is never reached because it would require an infinite amount of thrust and fuel. However it allows very, very large velocities to be reached making starship and uniship travel in reasonable time frames possible.

If the conditions in the quoted section 3.3 in the previous section of this book, then it is possible to reach "infinite" speed in a finite time in the earth's reference frame. To reach this limiting speed will require an infinite amount of starship time, and fuel.

# 4

# Baryonic Force Mechanism for Trips into the Multiverse

## *4.1 The Characteristics of a Rapidly Spinning Neutron Star*

It is estimated that there are 100 million neutron stars in our galaxy. It is likely that other galaxies of comparable size have a similar plethora of neutron stars. Neutron stars represent the end point of the evolution of main sequence stars with initial masses above ten solar masses. If such a star evolves and declines to a mass between the Chandrasekhar Limit (1.44 solar masses) and three solar masses, then it will become a neutron star – a star composed primarily of neutrons compacted to a radius of approximately ten kilometers with an enormous density of the order of $10^{17}$ kg/m$^3$. We expect neutron stars to be plentiful in our universe and other mature universes of the multiverse.

Neutron stars are thought to have a series of layers: an outer crust composed of a crust of iron with an estimated density of the order of $10^9$ kg/m$^3$ and an innermost purely neutron (possibly a quark-gluon plasma) core region with a density of order of $10^{18}$ kg/m$^3$ or more.

An important factor in the nature of neutron stars is their unusually large rotation rates ranging up to over 700 rotations per second. The rotation of a neutron star (assuming a baryonic gauge force exists as we do) generates a baryonic magnetic field as well as a baryonic electric field. However we shall see that the neutron star spin-generated baryonic magnetic and electric fields only impart a force to a uniship within our universe. Thus the Coulomb force is the key to slingshoting a uniship into the Flatverse as we pointed out in volume I.

The important neutron star parameters for the determination of the slingshot trajectory into the Flatverse are its mass, its density as function of radius, and its spin. From these parameters we can determine its baryon number, its baryonic current density, and its baryonic electric and magnetic fields. Then the trajectory of a uniship into the Flatverse can be determined.

## 4.2 The Baryonic Fields of a Rapidly Spinning Neutron Star

Spinning neutron stars, and they all spin at varying rates except in extreme old age, generate baryonic fields which *might* have provided a mechanism to escape from our universe into the multiverse.[27] In fact, we shall see that *the spin generated force does not* have a component in the direction of the Flatverse spatial coordinates. Thus they do not exit uniships into the Flatverse.

The baryonic Coulomb force does have components in Flatverse directions outside our universe and can slingshot uniships into the Flatverse. It provides the slingshot mechanism.

This section calculates the spin-dependent baryonic fields for a neutron star of mass M, internal mass density $\rho(r)$, radius R, volume V, and rotation rate $\Omega$ measured in rotations per second. We assume the neutron star is rigid and rotates uniformly. We believe this is a reasonable assumption in view of the close packing of the neutron star throughout most of its body.

### 4.2.1 Baryonic Current of a Neutron Star

We assume that a neutron star is to good approximation spherical although a rapidly spinning neutron star will deviate slightly from a sphere. We also assume that the neutron star can be treated as point-like since its radius of the order of 10 km is very small compared to the closest point of approach of a uniship which will be, at minimum, tens of thousands of kilometers distant.

The baryonic current of a spinning neutron star, if the coordinate system is oriented so that it spins around the z-axis, yields a current $J_\phi$ in the $\phi$ direction of spherical coordinates (r, θ, φ). The current is to good approximation

$$J_\phi = \int dr d\theta d\phi r^2 \beta_B \Omega \rho(r)/m_n = (\beta_B \Omega/m_n) \int dr d\theta d\phi r^2 \rho(r) = \beta_B \Omega M/m_n \qquad (4.1)$$

---

where $\beta_B$ is the baryonic charge (analogous to e in electromagnetism), where $m_n$ is the mass of a neutron, $\rho(r)/m_n$ is the neutron (nucleon) number density at radius r, and where the current at a radial distance r is $\beta_B \Omega \rho(r)/m_n$.

As discussed in volume I the baryonic potential in Flatverse coordinates is

$$\phi(y_1, y_2, \dots, y_{15}) = (\beta_B^2/4\pi)N_1 N_2/(y_1^2 + y_2^2 + \dots + y_{15}^2)^{\frac{1}{2}} \qquad (4.2)$$

where $\alpha_B = (\beta_B^2/4\pi)$ is the equivalent of the electromagnetic fine structure constant $\alpha$. In section 1.3 we estimated the order of magnitude of $\alpha_B$ using the not very well understood differences[28] between various experiments to determine the gravitational constant G. We found the order of magnitude to be

$$\alpha_B = \beta_B^2/4\pi \simeq .118\, Gm_H^2 \qquad (4.3)$$

where $m_H$ is the mass of a hydrogen atom. We will discuss the slingshot mechanism due to the baryonic Coulomb force in section 4.3.

We conclude this subsection by determining the baryonic current from the above equations:

$$J_\phi = (4\pi \cdot 0.118G)^{\frac{1}{2}}\Omega M \qquad (4.4)$$
$$= 1.22\, G^{\frac{1}{2}}\Omega M$$

using the approximation $m_H = m_n$.

## 4.2.2 The Sixteen Component Baryonic Vector Potential

In chapter 5 of volume I we described some of the features of the baryonic vector potential, which we recapitulate here for the reader's convenience:

As in electromagnetism there is an antisymmetric tensor of the second rank that appears in the free part of the baryonic field $F_{Bu_{\mu\nu}}(y)$ lagrangian:[29]

---

[28] The recent experiment by T. Quinn et al, Phys. Rev. Lett. **111**, 101102 (2013) differs significantly from the 2010 CODATA world average of previous experiments. See P. J. Mohr, B.N. Taylor, and D. B. Newell, Rev. Mod. Phys. **84**, 1527 (2012).. We attribute the difference to the baryonic force between masses.

[29] Parts of the following appear in Blaha (2014). They are somewhat modified since we are dealing with the classical, low energy, large distance baryonic field where the quantum coordinate fields Y(y) are well approxim ated by the classical (non-quantum) Flatverse coordinates y.

$$\mathcal{L}_{Bu} = -¼\, F_{Bu}{}^{ij}(y)F_{Buij}(y) \tag{5.1}$$

where

$$F_{Buij}(y) = \partial B_{ui}(y)/\partial y^j - \partial B_{uj}(y)/\partial y^i \tag{5.2}$$

and i, j = 1, 2, ... , 16. The $16^{th}$ coordinate corresponds to the time coordinate. While the coordinates are complex in general we will treat the 15 spatial coordinates as real and the $16^{th}$ coordinate as pure imaginary with the resulting invariant interval

$$ds^2 = dy_1{}^2 + dy_2{}^2 + ... + dy_{15}{}^2 - c^2 dy_{16}{}^2 \tag{5.3}$$

which is invariant under 16 dimensional Lorentz transformations.  The coordinates can be transformed into complex-valued coordinates using the Reality group defined in Blaha (2014) and earlier books.

The tensor $F_{Buij}$ is conveniently separated into an baryon electric part and a baryon magnetic part in a manner similar to the separation of the electromagnetic fields into electric and magnetic fields. However the 15 spatial dimensions change the forms of the baryon fields. Analogously, to electromagnetism the baryonic force is given by

$$f_i = F_{Buij}(y)J_B{}^j/c \tag{5.4}$$

where $J_B{}^j$ is the $j^{th}$ baryonic current.

The baryon "electric" field is

$$E_{Bui} = -F_{Bui0}(y)/c \tag{5.5}$$

while the baryon "magnetic" field is

$$B_{Bui} = \varepsilon_{ijk}F_{Bu}{}^{jk}(y) \tag{5.6}$$

where i, j, k = 1, 2, ... , 15 and where $\varepsilon_{ijk}$ is a totally anti-symmetric tensor with component values ±1. If i < j < k then $\varepsilon_{ijk}$ is +1. Even permutations of these three indices yield a value of +1 for the tensor components. Odd permutations of these three indices yield a value of −1. For example, $\varepsilon_{246} = +1$, $\varepsilon_{426} = -1$, $\varepsilon_{642} = -1$, $\varepsilon_{264} = -1$, $\varepsilon_{462} = +1$, $\varepsilon_{624} = +1$.

With these definitions of the $\mathbf{E_{Bu}}$ and $\mathbf{B_{Bu}}$ fields we derive the 16-dimensional generalization of the *Lorentz force law* for a baryon of charge q and 15-velocity $v_j$:

$$F_i = qE_{Bui} + q\varepsilon_{ijk}v_jB_{Buk}/c \tag{5.7}$$

for i = 1, 2, ... , 15. One important difference from the 4-dimensional case is the forms of the $\mathbf{E_{Bu}}$ and $\mathbf{B_{Bu}}$ fields

$$E_{Bui} = -F_{Bui0}(y)/c = [-\partial B_{u16}(y)/\partial y^i - \partial B_{ui}(y)/\partial y^{16}]/c \tag{5.8}$$

or, expressed as a 15-vector,

$$\mathbf{E_{Bu}} = [-\nabla_{15}\phi(y) - \partial\mathbf{B_u}(y)/\partial y^{16}]/c \tag{5.9}$$

where $\phi$ is the baryonic Coulomb potential $B_{u16}(y)$, $\nabla_{15}$ is the 15-dimensional grad operator, and $\mathbf{B_u}(y)$ is the baryonic 15-vector potential.

The 15-dimensional baryon magnetic field has the form of eqn. 5.6. A specific illustrative case shows the baryon magnetic field exhibits more complexity than the 3-dimensional magnetic field of electromagnetism:

$$B_{Bu1} = \varepsilon_{1jk}F_{Bu}{}^{jk}(y)/c = [F_{Bu}{}^{23}(y) + F_{Bu}{}^{24}(y) + \dots + F_{Bu}{}^{2,15}(y) + F_{Bu}{}^{34}(y) + F_{Bu}{}^{35}(y) + \dots + F_{Bu}{}^{3,15}(y) + F_{Bu}{}^{45}(y) + \dots + F_{Bu}{}^{14,15}(y)]/c \tag{5.10}$$

### 4.2.3 The Baryonic Electric and Magnetic Field Strengths

In this section we will calculate the baryonic electric and magnetic field strengths for a neutron star due to its spin. The spatial field strengths are determined by the dynamical equation:

$$\partial F_{Bu}{}^{ij}(y)/\partial y^i = J^j \tag{4.5}$$

By eq. 4.1 above the current for a neutron star is well approximated by the constant current in the $\phi$ direction (in the spherical coordinates)

$$J_\phi = \beta_B\Omega M/m_n \tag{4.1}$$

due to the neutron star's small size. Expressing $J_\phi$ in rectangular coordinates assuming the rotation is along the z-axis we find

$$J_x = -\sin\phi\, J_\phi \tag{4.6}$$

$$J_y = \cos\phi\, J_\phi \tag{4.7}$$

We will use the relative flatness of space, and the small size of the neutron star neighborhood, to identify the x, y, and z of our universe with the Flatverse coordinates $y^1$, $y^1$, and $y^3$. Inserting eqns. 4.6 and 4.7 in eq. 4.5 yields fifteen equations:

$$\partial F_{Bu}{}^{ix}(y)/\partial y^i = -\sin\phi\, J_\phi \tag{4.8}$$
$$\partial F_{Bu}{}^{iy}(y)/\partial y^i = \cos\phi\, J_\phi$$

$$\partial F_{Bu}{}^{ij}(y)/\partial y^j = 0$$

for j = 4, … , 15. These 15 equations have a solution that gives a magnetic force that is solely within the three spatial dimensions of our universe. *Thus they do not participate in the slingshot into the Flatverse.*[30]

## 4.3 The Baryonic Coulomb Force Slingshot into the Flatverse

The 16-dimensional version of the Lorentz force is

$$F_i = qE_{Bui} + q\varepsilon_{ijk}v_j B_{Buk}/c \qquad (4.9)$$

where

$$\mathbf{E_{Bu}} = [-\nabla_{15}\phi(y) - \partial \mathbf{B_u}(y)/\partial y^{16}]/c \qquad (4.10)$$

We have seen that the baryonic force slingshot is wholly derived from the baryonic Coulomb force (eq. 4.2),

$$\phi(y_1, y_2, … , y_{15}) = (\beta_B{}^2/4\pi)N_1 N_2/(y_1{}^2 + y_2{}^2 + … + y_{15}{}^2)^{\frac{1}{2}} \qquad (4.2)$$

between two baryon masses with baryon numbers $N_1$ and $N_2$. The baryonic Coulomb force is (eq. 5.20 of volume I):

$$F_i = \nabla_{15i}\phi(y) \qquad (5.20)$$

where $\nabla_{15i}$ is the $i^{th}$ component of the 15-dimensional grad operator $\nabla_{15}$.

The part of the Lorentz force that slingshots a uniship into the Flatverse is

$$F_{isling} = \partial\phi(y)/\partial y^i \qquad (4.11)$$
$$= (\beta_B{}^2/4\pi)N_1 N_2\, y_i\,/(y_1{}^2 + y_2{}^2 + … + y_{15}{}^2)^{3/2}$$

for i = 4, 5, … , 15.

A uniship will have a baryonic force directing it out of our universe into the Flatverse. The baryonic force between the uniship and the neutron star is repulsive since they are both composed of a majority of baryons.

---

[30] This discussion corrects the initial view of volume 1 that spin generated baryonic force can propel uniships into the Flatverse.

This force will undoubtedly be small compared to the gravitational forces during a slingshot maneuver. Thus we can see that the uniship will slowly ease out of our universe. For a time it will be partly in and partly out of the universe creating a physical situation not hitherto encountered in physics. We considered this sort of situation in section 2.5.3 previously. It has some advantages since, for example, it allows an "umbrella" of thrust tubes that originally are in 3-dimensional space to widen into a 15-dimensional umbrella of thrust tubes that would enable the uniship to travel in any direction in the Flatverse – Flatverse maneuverability. We shall consider this possibility in chapter 5.

Since the force of gravity is confined to within our universe it will have no effect on directions outside our universe. Baryonic forces in directions within our universe will be of little consequence compared to gravitation.

In directions into the Flatverse, gravitation forces being absent, the baryonic force will be the sole force. See chapter 5 for a detailed derivation.

Thus we find a clear division: gravitation dominates in directions within the universe; baryonic force dominates in directions out of our universe. Fig. 4.1 depicts the trajectory of a uniship in a slingshot maneuver. Note the attractive hyperbolic motion due to the dominance of gravity in our universe.

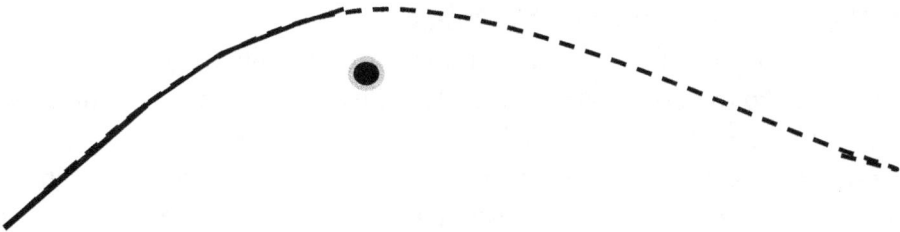

**Figure 4.1. The trajectory of a uniship in a slingshot maneuver with a neutron star (the dark circle). The repulsive baryonic force causes the "turn" away from the star into the Flatverse. The solid line corresponds to the time in which the uniship is wholly within the universe and dominated by gravity. The dotted line reflects the transition of the uniship into the Flatverse.**

# 5

# Point-like Uniship Slingshots

The major problems of a close approach of perhaps 100,000 km to a neutron star are the large gravity of the neutron star, its strong tidal gravitation effects (stresses) on the uniship's structure, its strong magnetic field, and its emission of large amount of primarily x-ray radiation. These properties of a neutron star neighborhood would appear to significantly affect the structural integrity of the uniship, and, more importantly, seriously impact on the safety and life of its human crew.

Fortunately there is a saving grace in this physical environment. A uniship approaching the neutron star could resolve these issues by traveling at extremely high speed[31] so that the time spent in the "danger" zone of the neutron star would be very small thus sharply reducing its deleterious effects.

## 5.1 The Uniship Slingshot Trajectory

The slingshot trajectory of a uniship is approximately a hyperbola in our universe due to the dominance of gravity. As it approaches a neutron star, with a distance of closest approach of perhaps 100,000 km, a uniship could be programed to take perhaps 3 seconds or so to circle around the neutron star. Consequently the uniship need only spend a minimal time near the neutron star with little gravitational tidal stress, magnetic field exposure, and radiation exposure issues.

Since the difficulties of a neutron star slingshot are surmountable, we will now turn to the issue of a uniship escape from our universe. We are fortunate that the neutron star's force on the uniship can be conveniently divided into two parts: gravitation which only influences the spatial motion of the uniship in our universe's coordinates, and the baryonic force which only significantly influences

---

[31] The starship would approach at perhaps c/2 and acquire an additional speed due to gravity of c/3. Combining these speeds using the rules of special relativity for the addition of velocities yields an approach speed of 0.7c near the neutron star.

the uniship's other 12 spatial Flatverse coordinates. The weakness of the baryonic force compared to gravity makes its impact on motion in our universe's spatial coordinates negligible.

The baryonic force generated by the interaction of a uniship's baryons with the baryons within the neutron universe causes the uniship's course to be deflected into the Flatverse.

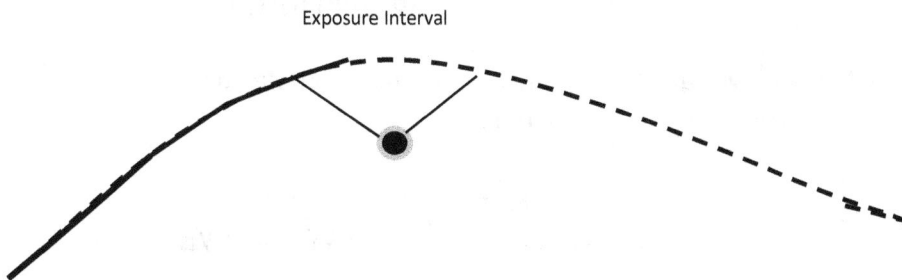

Exposure Interval

**Figure 5.1. Depiction of the uniship exposure interval of closest approach to a neutron star. The time interval spent in this region might be about three second or so. The uniship speed as it approaches might be about 0.7c or so to avoid capture by the neutron star. When the uniship totally exits our universe it disappears from view since electromagnetic radiation (light) from the uniship (or any object) "cannot penetrate" the boundary of our universe.[32] The dashed line indicates the partial exit of the uniship from our universe into the Flatverse with Flatverse velocity and momentum components.**

## 5.2 Uniship Neutron Star Slingshot Dynamics

In this section we will describe the dynamics of the neutron star slingshot. We will assume a flat space-time in our universe with the understanding that space-time may be significantly curved in the immediate vicinity of the neutron star. We assume the uniship will not enter that region. We will define the neutron star to be at the origin of the Flatverse coordinates. Thus $y_i = 0$ for $i = 1, \ldots, 15$. We will assume the universe spatial coordinates $x_i$ to be equal to the first three Flatverse coordinates:

---

[32] Cannot penetrate in the sense that it would have to travel an enormous distance from the edge of the universe in universe coordinates. This topic is discussed in section 2.4.8.

$$y_i = x_i \tag{5.1}$$

for i = 1, 2, 3. We are allowed to do this by the flat space-time assumption for the universe.

The total potential energy of the uniship in the neutron star's reference frame is[33]

$$V_{tot} = -GM_1M_2/r_u + (\beta_B^2/4\pi)\, N_1N_2/r_F \tag{5.2}$$

where M1 and M2 are the masses of the neutron star and uniship, N1 and N2 are their baryon numbers, and where

$$r_u^2 = x_1^2 + x_2^2 + x_3^2 = y_1^2 + y_2^2 + y_3^2 \tag{5.3}$$
$$r_F^2 = y_1^2 + \dots + y_{15}^2 = r_u^2 + y_4^2 + \dots + y_{15}^2 \tag{5.4}$$

The force is the gradient of the potential

$$
\begin{aligned}
F^i_{slingshot} &= \partial\, V_{tot}/\partial y_i \\
&= GM_1M_2\, (\delta^{i1} + \delta^{i2} + \delta^{i3})y^i/r_u^{3/2} - (\beta_B^2/4\pi)N_1N_2\, y^i/r_F^{3/2}
\end{aligned} \tag{5.5}
$$

where the Kronecker delta functions restrict the gravity force to the spatial coordinates of the universe.

The dynamic equation of the uniship motion[34] is

$$dp^i/d\tau = F^i_{slingshot} \tag{5.6}$$

where $\tau$ is the invariant interval. We now consider the initial phase of the escape to the Flatverse in which

$$r_u \gg r_F - r_u \tag{5.7}$$

The universe spatial distance is much greater than the purely Flatverse spatial distance $r_F - r_u$. In this case eq. 5.6 has different forms to good approximation for i = 1, 2, 3 and the remaining Flatverse spatial coordinates:

---

[33] Again we note that we are assuming a uniship trajectory in flat space-time so that we may use special relativistic potentials and dynamic equations.
[34] The neutron star is assumed to be stationary due to the largeness of its mass relative to the uniship.

$$dp^i/d\tau \simeq GM_1M_2y^i/r_u^{3/2} \tag{5.8}$$

for i = 1, 2, 3 and

$$dp^i/d\tau = -(\beta_B^2/4\pi)N_1N_2\, y^i/r_F^{3/2} \tag{5.9}$$

for i = 4, ... , 15. Eq. 5.8 yields a solution of the central force problem which in the present case is an approximately hyperbolic trajectory of the form depicted in Fig. 5.1.

Due to our well-justified assumption that the distance into the Flatverse will be small compared to the distance of the uniship from the neutron star we can approximate

$$r_F^{-3/2} \simeq r_u^{-3/2}[1 - (3/2)(r_F^2 - r_u^2)/r_u^2] \tag{5.10}$$

Eq. 5.9 then becomes approximately

$$dp^i/d\tau \simeq -(\beta_B^2/4\pi)N_1N_2\, y^i[1 - (3/2)(r_F^2 - r_u^2)/r_u^2]/r_u^{3/2} \tag{5.11}$$

or

$$dp^i/d\tau \simeq -(\beta_B^2/4\pi)N_1N_2\, y^i[1 - (3/2)\Sigma y_k^2/r_u^2]/r_u^{3/2} \tag{5.12}$$

where the sum in eq. 5.12 is from k = 4, ..., 15. Eqns. 5.11 and 5.12 yields an initially exponential-like trajectory into the Flatverse to leading order.

Thus we have shown that the neutron star slingshot clearly drives the uniship out of the universe into the Flatverse. The uniship passes into the Flatverse. For a small time it is partly in and partly out of the universe as it passes through the universe's horizon. This situation is described in section 2.4.7.

# 6

# Umbrella-Shaped Uniship Slingshots

We take it for granted that we can move in one spatial direction or another with ease. However when one enters a higher dimensional space from a space of lower dimension, movement in the additional dimensions, which requires the expenditure of force in those dimensions, becomes an issue. A lower dimension object does not automatically have forces within it in the additional directions and so it cannot move itself, or part of itself in those directions.

In the previous chapter we saw how to use the baryonic force to escape from our universe to the higher dimension Flatverse. In this chapter we will show how to give "wings" to a uniship so that it will have the ability to maneuver in any direction in the Flatverse. To achieve this capability we will have to design the uniship so that it will expand in all Flatverse directions as it enters the Flatverse. This will require an umbrella-like configuration that will use the baryonic force to open the umbrella in all Flatverse directions. The spokes of the umbrella will be long thrust tubes through which uniship thrust can be directed to move the uniship in a desired direction. Fig. 6.1 is a depiction of the simplest form of umbrella uniship.

## 6.1 Scenario for the Opening of a Uniship Umbrella

As an umbrella uniship slingshots around a neutron star the body of the ship including fuel tanks is rigid and moves as a unit. The spokes of the umbrella, the thrust tubes, are moveable and will each move differently because of their differing average distance from the neutron star. As a result they will point in different directions in the Flatverse and can be further moved within the Flatverse relative to each other to deliver thrust in any Flatverse direction. After positioning they can be locked in place and used to maneuver the uniship towards any universe. (Flatverse havigation is discussed in volume I.)

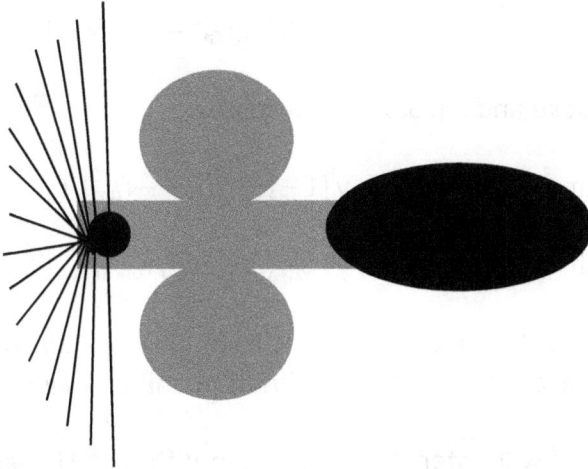

**Figure 6.1. Tentative umbrella-like uniship design with the spokes of the umbrella forming a fan. The thrust tubes (umbrella spokes) extend kilometers from the thrust power generator(s) core to enable the baryonic force to maneuver them in the 15 different Flatverse directions. The thrust tubes are able to swivel into all 15 spatial directions in response to the baryonic force as the uniship enters the Flatverse. Two fuel spheres are depicted under the assumption that one holds hydrogen and the other holds anti-hydrogen since they are presently the most powerful known possible energy source. The black forward part is for crew, supplies, and cargo. The rear gray part holds the engine apparatus and other related engine components.**

## 6.2 Equations for the Motion of the Thrust Tubes Entering the Flatverse

In chapter 5 we developed the equations for the slingshot mechanism for a compact rigid uniship. In this section we will extend the equations to the case of a rigid uniship with an umbrella with a fan shape (a flat array of spokes as picture in Fig. 6.1. The array is flat with each spoke in the plane of the ship's trajectory.) A uniship with a true umbrella of spokes (thrust tubes) is another possibility that may be of importance. This is a technical question that we will not address.

We define radial distances from the neutron star center for a "fan-shaped" umbrella with the n[th] spoke end[35] at radial distance $r_{un}$. Then eq. 5.10 becomes

$$r_{Fn}{}^{-3/2} \simeq r_{un}{}^{-3/2}[1 - (3/2)(r_{Fn}{}^2 - r_{un}{}^2)/r_{un}{}^2] \qquad (6.1)$$

for the end of each spoke and eq. 5.9 then becomes

$$dp_n{}^i/d\tau \simeq -(\beta_B{}^2/4\pi)N_1 N_2\, y^i[1 - (3/2)(r_{Fn}{}^2 - r_{un}{}^2)/r_{un}{}^2]/r_{un}{}^{3/2} \qquad (6.2)$$

or

$$dp_n{}^i/d\tau \simeq -(\beta_B{}^2/4\pi)N_1 N_2\, y^i[1 - (3/2)\Sigma y_k{}^2/r_{un}{}^2]/r_{un}{}^{3/2} \qquad (6.3)$$

for the n[th] spoke where the sum in eq. 6.3 is from k = 4, …, 15. For each spoke radial distance $r_{un}$ these equations yield different initial trajectories into the Flatverse to leading order.

Thus the spokes will enter the Flatverse in different Flatverse directions since they are not rigidly attached to the uniship body. Upon entry they can change each other's direction giving a final configuration with full Flatverse maneuverability. Uniships can now move in any Flatverse spatial directions.

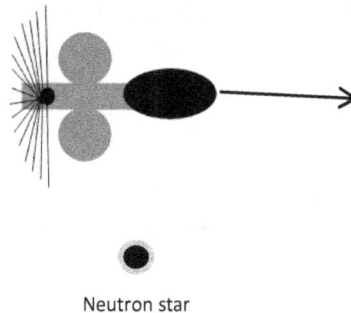

Neutron star

**Figure 6.2. Uniship slingshot past neutron star. Note the umbrella spokes which are attached to the uniship but moveable will respond differently to the baryonic force since they are at different radial distances from the neutron star and thus feel differing amounts of force. The baryonic force will twist them in different directions. They then can be re-oriented by the uniship computer to provide mobility in all Flatverse directions.**

---

[35] The spoke center of mass actually.

# 7

# The Dimensions of the Parts of a Uniship in the Flatverse

A uniship in the Flatverse has parts with different dimensionality. Some parts are 3-dimensional; some parts are 15-dimensional. We will examine the dimensionality of the various uniship parts in this chapter. Fig. 7.1 displays the parts of a uniship crudely mapped to two dimensions.

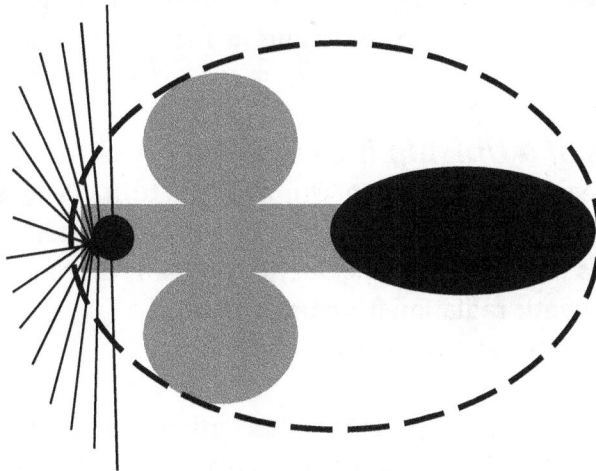

**Figure 7.1. A 2-dimensional projection of a uniship with the actual dimensions of each part in the Flatverse marked. The dashed ellipse encloses the 3-dimensional part of the uniship. The white space within the ellipse is 15-dimensional Flatverse space. The thrust tubes (spokes) are partly in the 3-dimensional uniship bubble. The outer parts of the thrust tubes are in the 15-dimensional Flatverse space pointing in the various Flatverse directions.**

## 7.1 Dimension of a Uniship in the Flatverse

From an overall perspective, a uniship is a 3-dimensional irregularly-shaped bubble in the 15-dimensional Flatverse space. We take the time dimension to be the same in our 4-dimensional universe (and thus the uniship) and in the 16-dimensional Flatverse space-time. This is the simplest choice of times. More complex choices of time in each case are possible. But the physics in any case will remain the same.

Since the space-time in the uniship is the same as our space-time the physical processes will be the same as they would be in our universe.

The thrust tubes are somewhat different. They are presumably long – of the order of kilometers so that the slingshot into the Flatverse will swing them in the fifteen spatial directions of the Flatverse. However, being attached to the 3-dimensional uniship they will be 3-dimensional at their beginning at the uniship end, and transition to 15-dimensional beyond that point. Despite this change of dimension they will each still be linear –not curved. A simple geometric example of this is a line attached to a circle's edge – all in the x-y plane. If the end of the line is swung through some means into the z direction, then one end will be in the x-y plane and the other end pointing in the z direction. The line remains straight.

## 7.2 Viewports of a Uniship into the Flatverse

Seeing is easy in our universe. Our eyes can turn in any direction and they are naturally adapted to seeing a range of electromagnetic radiation. In the Flatverse seeing is more problematic due to the many more dimensions, and the lack of electromagnetic radiation from universes.

A solution to these problems is:

1.  To "see" using baryonic radiation using devices that manipulate and transform baryonic radiation into visible radiation displayed on monitors.
2.  To sense baryonic radiation with detectors positioned on the end of thrust tubes. Each detector will detect baryonic radiation from all directions through a controllable rotating 15-dimensional mechanism. The radiation will be transformed into electromagnetic signals transmitted down a cable to the uniship body. There they will be combined like insects combine light from their eye facets into

composite images that can be viewed and used for navigation by the crew. Thus the thrust tubes have a dual function: to drive the uniship and to gather baryonic radiation so that the crew can view the universes of the multiverse.

Thus it is feasible to navigate in the multiverse using baryonic radiation. There are significant technical hurdles such as the creation of small baryonic radiation detection devices, and the creation of the equivalent of the mechanisms that manipulate and magnify electromagnetic radiation.

# 8

# Exploring Physical Infinity – The Final Step

We have viewed the skies of our universe for countless millennia. Eventually we will have explored the interesting parts of the universe and colonized the more pleasant parts.

Then the challenge of the multiverse will arise. Restless Man will seek to find new phenomena – new experiences – so that Man's search for a complete understanding of reality and consciousness can continue knowing that the goal is truly unreachable, and yet the quest is the key endeavor that makes Man worthy to exist.

# REFERENCES

Blaha, S., 1998, *Cosmos and Consciousness* (Pingree-Hill Publishing, Auburn, NH, 1998).

_____2004, *Quantum Big Bang Cosmology: Complex Space-time General Relativity, Quantum Coordinates™ Dodecahedral Universe, Inflation, and New Spin 0, ½, 1 & 2 Tachyons & Imagyons*  (Pingree-Hill Publishing, Auburn, NH, 2004).

_____ 2005a, *Quantum Theory of the Third Kind: A New Type of Divergence-free Quantum Field Theory Supporting a Unified Standard Model of Elementary Particles and Quantum Gravity based on a New Method in the Calculus of Variations* (Pingree-Hill Publishing, Auburn, NH, 2005).

_____, 2005b, *The Metatheory of Physics Theories, and the Theory of Everything as a Quantum Computer Language* (Pingree-Hill Publishing, Auburn, NH, 2005).

_____, 2005c, *The Equivalence of Elementary Particle Theories and Computer Languages: Quantum Computers, Turing Machines, Standard Model, Superstring Theory, and a Proof that Gödel's Theorem Implies Nature Must Be Quantum* (Pingree-Hill Publishing, Auburn, NH, 2005).

_____, 2006, *A Derivation of ElectroWeak Theory based on an Extension of Special Relativity; Black Hole Tachyons; & Tachyons of Any Spin*. (Pingree-Hill Publishing, Auburn, NH, 2006).

_____, 2007b, *The Origin of the Standard Model: The Genesis of Four Quark and Lepton Species, Parity Violation, the ElectroWeak Sector, Color SU(3), Three Visible Generations of Fermions, and One Generation of Dark Matter with Dark Energy* (Pingree-Hill Publishing, Auburn, NH, 2007).

_____, 2008a, *A Direct Derivation of the Form of the Standard Model From GL(16) (Pingree-Hill Publishing, Auburn, NH, 2008).*

_____, 2008b, *A Complete Derivation of the Form of the Standard Model With a New Method to Generate Particle Masses  Second Edition* (Pingree-Hill Publishing, Auburn, NH, 2008)

_____, 2009a, *Bright Stars, Bright Universe* (Pingree-Hill Publishing, Auburn, NH, 2009)

_____, 2009b, *To Far Stars and Galaxies: Second Edition of Bright Stars, Bright Universe*  (Pingree-Hill Publishing, Auburn, NH, 2009).

_____, 2009c, *The Algebra of Thought & Reality: The Mathematical Basis for Plato's Theory of Ideas, and Reality Extended to Include A Priori Observers and Space-Time Second Edition* (Pingree-Hill Publishing, Auburn, NH, 2009).

_____, 2010a, *Operator Metaphysics: A New Metaphysics Based on a New Operator Logic and a New Quantum Operator Logic that Lead to a Mathematical Basis for Plato's Theory of Ideas and Reality* (Pingree-Hill Publishing, Auburn, NH, 2010).

_____, 2010b, *The Standard Model's Form Derived from Operator Logic, Superluminal Transformations and GL(16)* (Pingree-Hill Publishing, Auburn, NH, 2010).

_____, 2011a, *$21^{st}$ Century Natural Philosophy of Ultimate Physical Reality* (McMann-Fisher Publishing, Auburn, NH, 2011).

_____, 2011b, *All the Universe! Faster Than Light Tachyon Quark Starships & Particle Accelerators with the LHC as a Prototype Starship Drive Scientific Edition* (Pingree-Hill Publishing, Auburn, NH, 2011).

_____, 2011c, *From Asynchronous Logic to The Standard Model to Superflight to the Stars* (Blaha Research, Auburn, NH, 2011).

_____, 2012a, *From Asynchronous Logic to The Standard Model to Superflight to the Stars volume 2: Superluminal CP and CPT, U(4) Complex General Relativity and The Standard Model, Complex Vierbein General Relativity, Kinetic Theory, Thermodynamics* (Blaha Research, Auburn, NH, 2012).

_____, 2012b, *Standard Model Symmetries, And Four And Sixteen Dimension Complex Relativity; The Origin Of Higgs Mass Terms* (Blaha Research, Auburn, NH, 2012).

_____, 2013a, *Multi-Stage Space Guns, Micro-Pulse Nuclear Rockets, and Faster-Than-Light Quark-Gluon Ion Drive Starships* (Blaha Research, Auburn, NH, 2013).

_____, 2013b, *The Bridge to Dark Matter; A New Sister Universe; Dark Energy; Inflatons; Quantum Big Bang; Superluminal Physics; An Extended Standard Model Based on Geometry* (Blaha Research, Auburn, NH, 2013).

_____, 2014a, *Universes and Multiverses: From a New Standard Model to a Physical Multiverse; The Big Bang, and Our Sister Universe Wormhole; Cosmological Constant's Origin; A Baryonic Field between Universes and*

*Particles; Flatverse Extended Wheeler-DeWitt Equation* (Blaha Research, Auburn, NH, 2014).

————, 2014b, *All the Multiverse! Starships Exploring the Endless Universes of the Cosmos using the Baryonic Force* (Blaha Research, Auburn, NH, 2014)

# INDEX

# About the Author

Stephen Blaha is an internationally known physicist with interests in Science, the Arts, and Technology. He had an Alfred P. Sloan Foundation scholarship in college. He received his Ph.D. in Physics from Rockefeller University. He has served on the faculties of several major universities. He was also a Member of the Technical Staff at Bell Laboratories, a manager at the Boston Globe Newspaper, a Director at Wang Laboratories, and President of Blaha Software Inc and of Janus Associates Inc. (NH).

Among other achievements he was a co-discoverer of the "r potential" for heavy quark binding developing the first (and still the only demonstrable) non-abelian gauge theory with an "r" potential; first suggested the existence of topological structures in superfluid He-3; first proposed Yang-Mills theories would appear in condensed matter phenomena with non-scalar order parameters; first developed a grammar-based formalism for quantum computers and applied it to elementary particle theories; first developed a new form of quantum field theory without divergences (thus solving a major 60 year old problem that enabled a unified theory of the Standard Model and Quantum Gravity without divergences to be developed); first developed a formulation of complex General Relativity based on analytic continuation from real space-time; first developed a generalized non-homogeneous Robertson-Walker metric that enabled a quantum theory of the Big Bang to be developed without singularities at t = 0; first generalized Cauchy's theorem and Gauss' theorem to complex, curved multi-dimensional spaces; received Honorable Mention in the Gravity Research Foundation Essay Competition in 1978; first developed a physically acceptable theory of faster-than-light particles; first showed a universe with three complex spatial dimensions is icosahedral; first derived a composition of extrema method in the Calculus of Variations; first quantitatively suggested that inflationary periods in the history of the universe were not needed; first proved Gödel's Theorem implies Nature must be quantum; provided a new alternative to the Higgs Mechanism, and Higgs particles, to generate masses; first showed how to resolve logical paradoxes including Gödel's Undecidability Theorem by developing Operator Logic and Quantum Operator Logic; first developed a quantitative harmonic oscillator-like model of the life cycle, and interactions, of civilizations; first showed how equations describing superorganisms also apply to civilizations; and first developed an axiomatic derivation of the forms of The Standard Model with DARK PARTICLEs from geometry – space-time properties – The faster than light Standard Model.

He has had a major impact on a succession of elementary particle theories: his Ph.D. thesis (1970), and papers, showed that quantum field theory calculations to all orders in ladder approximations could not give scaling deep inelastic electron-nucleon scattering. He later showed the eigenvalue equation for the fine structure constant $\alpha$ in Johnson-Baker-Willey QED had a zero at $\alpha = 1$ not $1/137$ by solving the Schwinger-Dyson equations to all orders in an approximation that agreed with exact results to $8^{th}$ order in $\alpha$ thus ending interest in this theory. In 1979 at Prof. Ken Johnson's (MIT) suggestion he calculated the proton-neutron mass difference in the MIT bag model and found the result had the wrong sign reducing interest in the bag model. These results all appear in Physical Review papers. In the 2000's he repeatedly pointed out the shortcomings of SuperString theory and showed that The Standard Model's form could be derived from space-time geometry by an extension of space-time to complex-valued coordinates and the Lorentz group to the complex Lorentz group which supports faster than light transformations. This deeper space-time basis shows that the Extended Standard Model developed by Blaha has an origin in geometry and is the true theory of elementary particles.

More recently Blaha has developed a theory of the Multiverse based on a complex Euclidian 16-dimensional space that explains (using the Wheeler-DeWitt equation for quantum gravity generalized to complex-valued space-times) the origin of the Cosmological Constant, the origin for the spatial asymmetry of the Universe, and an understanding of the origin of the newly found Web of Galaxies (that links all the groups of galaxies) in our universe.

He also developed proposals for faster than light starships using quark-gluon ion drives and drives based on particle-antiparticle annihilation.

In the early 1980's Blaha was also a pioneer in the development of UNIX for financial, scientific and Internet applications: benchmarked UNIX versions showing that block size was critical for UNIX performance, developing financial modeling software, starting database benchmarking comparison studies, developing Internet-like UNIX networking (1982) and developing a hybrid shell programming technique (1982) that was a precursor to the PERL programming language. He was also the manager of the AT&T ten-year future products development database. His work helped lead to commercial UNIX on computers such as Sun Micros, IBM AIX minis, and Apple computers.

In the 1980's he pioneered the development of PC Desktop Publishing on laser printers. and was nominated for three "Awards for Technical Excellence" in 1987 by PC Magazine for PC software products that he designed and developed.

In the past twelve years Dr. Blaha has written over 40 books on a wide range of topics. Some recent major works are: *From Asynchronous Logic to The Standard Model to Superflight to the Stars, All the Universe!* and *SuperCivilizations: Civilizations as Superorganisms.*

www.ingramcontent.com/pod-product-compliance
Lightning Source LLC
Chambersburg PA
CBHW082112210326
41599CB00033B/6678